PHILIP'S BRITAIN & IRELAND

2020 STARGAZING

MONTH-BY-MONTH GUIDE TO THE NIGHT SKY

HEATHER COUPER & NIGEL HENBEST

www.philipsastronomy.com
www.philips-maps.co.uk

Published in Great Britain in 2019 by Philip's,
a division of Octopus Publishing Group Limited
(www.octopusbooks.co.uk)
Carmelite House, 50 Victoria Embankment,
London EC4Y 0DZ
An Hachette UK Company (www.hachette.co.uk)

ISBN 978-184907-520-6

A CIP catalogue record for this book is available from the British Library.

Printed in China

CONTENTS

Welcome to the latest *Stargazing*! Within these pages, you'll find your complete guide to everything that's happening in the night sky throughout 2020 – whether you're a beginner or an experienced astronomer.

This year, we're celebrating the amazing images that anyone can take from their backyard. Since *Stargazing* first came out in 2005, all our monthly pictures have been taken by British amateur astronomers. And check out Robin Scagell's article on pages 86–89 to find out how to do it yourself – even with your smartphone!

With the 12 monthly star charts, you can find your way around the sky on any night in the year. Impress your friends by identifying celestial sights ranging from the brightest planets to some pretty obscure constellations.

Every page of *Stargazing 2020* is bang up to date, bringing you everything that's new this year, from shooting stars to eclipses. And we'll start with a rundown of the most exciting sky-sights on view in 2020 (opposite).

THE MONTHLY CHARTS

A reliable map is just as essential for exploring the heavens as it is for visiting a foreign country. For each month, we provide a circular **star chart** showing the whole evening sky.

To keep the maps uncluttered, we've plotted about 200 of the brighter stars (down to third magnitude), which means you can pick out the main star patterns – the constellations. (If we'd shown every star visible on a really dark night, there'd be around 3000 stars on the charts!) We also show the ecliptic: the apparent path of the Sun in the sky, it's closely followed by the Moon and planets as well.

You can use these charts throughout the UK and Ireland, along with most of Europe, North America and northern Asia – between 40 and 60 degrees north – though our detailed timings apply specifically to the UK and Ireland.

USING THE STAR CHARTS

It's pretty easy to use the charts. Start by working out your compass points. South is where the Sun is highest in the sky during the day; east is roughly where the Sun rises, and west where it sets. At night, you can find north by locating the Pole Star – Polaris – by using the stars of the Plough (see April).

The left-hand chart then shows your view to the north. Most of the stars here are visible all year: these circumpolar constellations wheel around Polaris as the seasons progress.

Your view to the south appears in the right-hand chart; it changes much more as the Earth orbits the Sun. Leo's prominent 'sickle' is high in the spring skies. Summer is dominated by the bright trio of Vega, Deneb and Altair. Autumn's familiar marker is the Square of Pegasus, while the stars of Orion rule the winter sky.

During the night, our perspective on the sky also alters as the Earth spins around, making the stars and planets appear to rise in the east and set in the west. The charts depict the sky in the late evening (the exact times are noted in the captions). As a rule of thumb, if you are observing two hours later, then the following month's map will be a

HIGHLIGHTS OF THE YEAR

- **Night of 3/4 January:** the **Quadrantid meteor shower** is spectacular after midnight.
- **10 January:** the Full Moon looks tarnished as it suffers a penumbral eclipse.
- **27 January:** see the brightest and faintest planets together, as Venus grazes Neptune.
- **17–18 February:** Mars glides between the Lagoon Nebula and the Trifid Nebula.
- **27 February:** the crescent Moon and Venus form a beautiful close pair after sunset.
- **23 March (am):** Pluto lies immediately behind Mars (moderate telescope needed).
- **28 March:** a lovely tableau of Venus, the crescent Moon, the Pleiades and Aldebaran.
- **29 March:** the Moon occults the Hyades, with Venus nearby.
- **3 April:** Venus in front of the Pleiades.
- **8 April:** the closest and brightest supermoon of 2020.
- **Night of 21/22 April:** it's an excellent year for observing the **Lyrid meteor shower**.
- **26 April:** there's a gorgeous pairing of Venus and the crescent Moon after sunset.
- **21–22 May:** Mercury lies near Venus in the evening sky.
- **24 May:** you'll find Mercury between the narrow crescent Moon and Venus.
- **19 June (am):** the crescent Moon moves right in front of Venus (during daylight).
- **21 June:** an annular solar eclipse is visible from a narrow band crossing central Africa, northern India and China. Not visible from the British Isles.
- **5 July:** the Full Moon passes below Jupiter.
- **14 July:** Jupiter is opposite to the Sun in the sky and at its brightest this year.
- **18 July (am):** the crescent Moon moves in front of the Crab Nebula (telescope needed).
- **20 July:** Saturn is opposite to the Sun in the sky and at its brightest this year.
- **9 August (am):** use binoculars to spot Mercury in front of the Praesepe star cluster.
- **Night of 12/13 August:** maximum of the prolific **Perseid meteor shower**, but the show is spoilt this year by moonlight.
- **13 August (am):** a beautiful view of the crescent Moon in the Hyades, with the Pleiades and brilliant Venus.
- **Night of 5/6 September:** the Moon is very close to brilliant red Mars.
- **11 September:** Neptune is opposite to the Sun in the sky and at its brightest this year.
- **3 October (am):** brilliant Venus passes so close to Regulus that it looks as though the star has gone supernova!
- **13 October:** Mars lies opposite to the Sun in the sky; the Red Planet is nearest to us on 6 October.
- **Night of 21/22 October:** the **Orionid meteor shower** is best observed after midnight, when the Moon has set.
- **31 October:** Uranus is opposite to the Sun in the sky and at its brightest this year.
- **Night of 13/14 December:** the bright slow shooting stars of the **Geminid meteor shower** provide the year's best celestial fireworks.
- **14 December:** a total eclipse of the Sun is visible from a narrow track starting in the South Pacific, crossing southern Chile and Argentina. Nothing is visible from the British Isles.
- **21 December:** Jupiter and Saturn are incredibly close together, and to the naked eye they seem to merge. It's their closest encounter since 1623.

better guide to the stars on view – though beware: the Moon and planets won't be in the right place!

THE PLANETS, MOON AND SPECIAL EVENTS

Our charts also highlight the **planets** above the horizon in the late evening. We've indicated the track of any **comets** known at the time of writing, though we're afraid we can't guide you to a comet that's found after the book has been printed!

We've plotted the position of the Full Moon each month, and also the **Moon's position** at three-day intervals before and afterwards. If there's a **meteor shower** in the month, we mark its radiant – the position from which the meteors stream.

The **Calendar** provides a daily guide to the Moon's phases and other celestial happenings. We've detailed the most interesting in the **Special Events** section, including close pairings of the planets, times of the equinoxes and solstices and **eclipses** of the Moon and Sun.

Check out the **Planet Watch** page for more about the other worlds of the Solar System. We've illustrated unusual planetary and lunar goings-on in the **Planet Event Charts**. And there's a full explanation of all these events in **Solar System 2020** on pages 80–82.

MONTHLY OBJECTS, TOPICS AND PICTURES

Each month, we examine one particularly interesting **object**: a planet, a star or a galaxy. We also feature a spectacular **picture** – taken from the UK or by a British amateur astronomer – describing how the image was captured, and subsequently processed to enhance the end result. And we explore a fascinating and often newsworthy **topic**, ranging from aurorae to black holes

GETTING IN DEEP

There's a practical **observing tip** each month, helping you to explore the sky with the naked eye, binoculars or a telescope.

Check out our guide to the **Top 20 Sky Sights**, such as nebulae, star clusters and galaxies. You'll find it on pages 83–85.

It's followed by equipment expert Robin Scagell's in-depth article on how to take the best astronomical photographs from your home patch.

If you're plagued by light pollution, use the dark-sky maps (pages 90–95). They show you where to find the blackest skies in Great Britain, and enjoy the most breathtaking views of the heavens.

So: fingers crossed for good weather, beautiful planets, a multitude of meteors and – the occasional surprise.

Happy stargazing!

JARGON BUSTER

Have you ever wondered how astronomers describe the brightness of the stars or how far apart they appear in the sky? Not to mention how we can measure the distances to the stars? If so, you can quickly find yourself mired in some arcane astro-speak – magnitudes, arcminutes, light years and the like.

Here's our quick and easy guide to busting that jargon:

Magnitudes

It only takes a glance at the sky to see that some stars are pretty brilliant, while many more are dim. But how do we describe to other people how bright a star appears?

Around 2000 years ago, ancient Greek astronomers ranked the stars into six classes, or **magnitudes**, depending on their brightness. The most brilliant stars were first magnitude, and the faintest stars you can see came in at sixth magnitude. So the stars of the Plough, for instance, are second magnitude while the individual Seven Sisters in the Pleiades are fourth magnitude.

Mars (magnitude -1.6 here) shines a hundred times brighter than the Seven Sisters in the Pleiades, which are around 5 magnitudes fainter.

Today, scientists can measure the light from the stars with amazing accuracy. (Mathematically speaking, a difference of five magnitudes represents a difference in brightness of one hundred times.) So the Pole Star is magnitude +2.0, while Rigel is magnitude +0.1. Because we've inherited the ancient ranking system, the brightest stars have the *smallest* magnitude. In fact, the brightest stars come in with a negative magnitude, including Sirius (magnitude -1.5).

And we can use the magnitude system to describe the brightness of other objects in the sky, such as stunning Venus, which can be almost as brilliant as magnitude -5. The Full Moon and the Sun have whopping negative magnitudes!

At the other end of the scale, stars, nebulae and galaxies with a magnitude fainter than +6.5 are too dim to be seen by the naked eye. Using ever larger telescopes – or by observing from above Earth's atmosphere – you can perceive fainter and fainter objects. The most distant galaxies visible to the Hubble Space Telescope are ten billion times fainter than the naked-eye limit.

Here's a guide to the magnitude of some interesting objects:

Sun	-26.7
Full Moon	-12.5
Venus (at its brightest)	-4.7
Sirius	-1.5
Betelgeuse	+0.4
Polaris (Pole Star)	+2.0
Faintest star visible to the naked eye	+6.5
Faintest star visible to the Hubble Space Telescope	+31

Degrees of separation

Astronomers measure the distance between objects in the sky in **degrees** (symbol °): all around the horizon is 360°, while it's 90° from the horizon to the point directly overhead (the zenith).

As we show in the photograph, you can use your hand – held at arm's length – to give a rough idea of angular distances in the sky.

For objects that are very close together

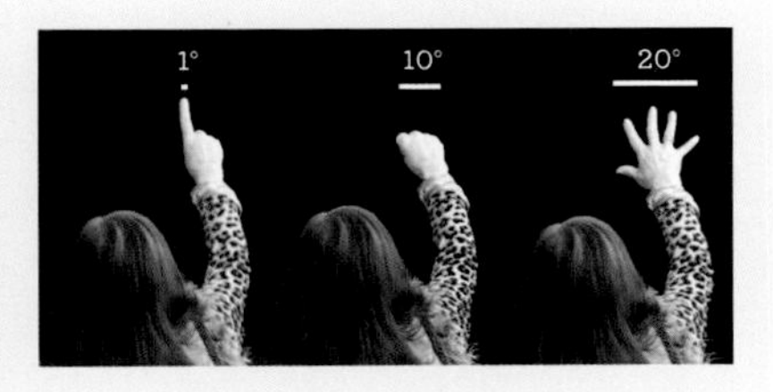

– like many double stars – we divide the degree into 60 arcminutes (symbol '). And for celestial objects that are very tiny – such as the discs of the planets – we split each arcminute into 60 arcseconds (symbol "). To give you an idea of how small these units are, it takes 3600 arcseconds to make up one degree.

Here are some typical separations and sizes in the sky:

Length of the Plough	25°
Width of Orion's Belt	3°
Diameter of the Moon	31'
Separation of Mizar and Alcor	12'
Diameter of Jupiter	45"
Separation of Albireo A and B	35"

How far's that star?

Everything we see in the heavens lies a long way off. We can give distances to the planets in millions of kilometres. But the stars are so distant that even the nearest, Proxima Centauri, lies some 40 million million kilometres away. To turn those distances into something more manageable, astronomers use a larger unit: one **light year** is the distance that light travels in a year.

One light year is about 9.46 million million kilometres. That makes Proxima Centauri a much more manageable 4.2 light years away from us. Here are the distances to some other familiar astronomical objects, in light years:

Sirius	8.6
Betelgeuse	720
Centre of the Milky Way	26,000
Andromeda Galaxy	2.5 million
Most distant galaxies seen by Hubble Space Telescope	13 billion

- The sky at 10 pm in mid-January, with Moon positions at three-day intervals either side of Full Moon.
- The star positions are also correct for 11 pm at the beginning of January, and 9 pm at the end of the month.
- The planets move slightly relative to the stars during the month.

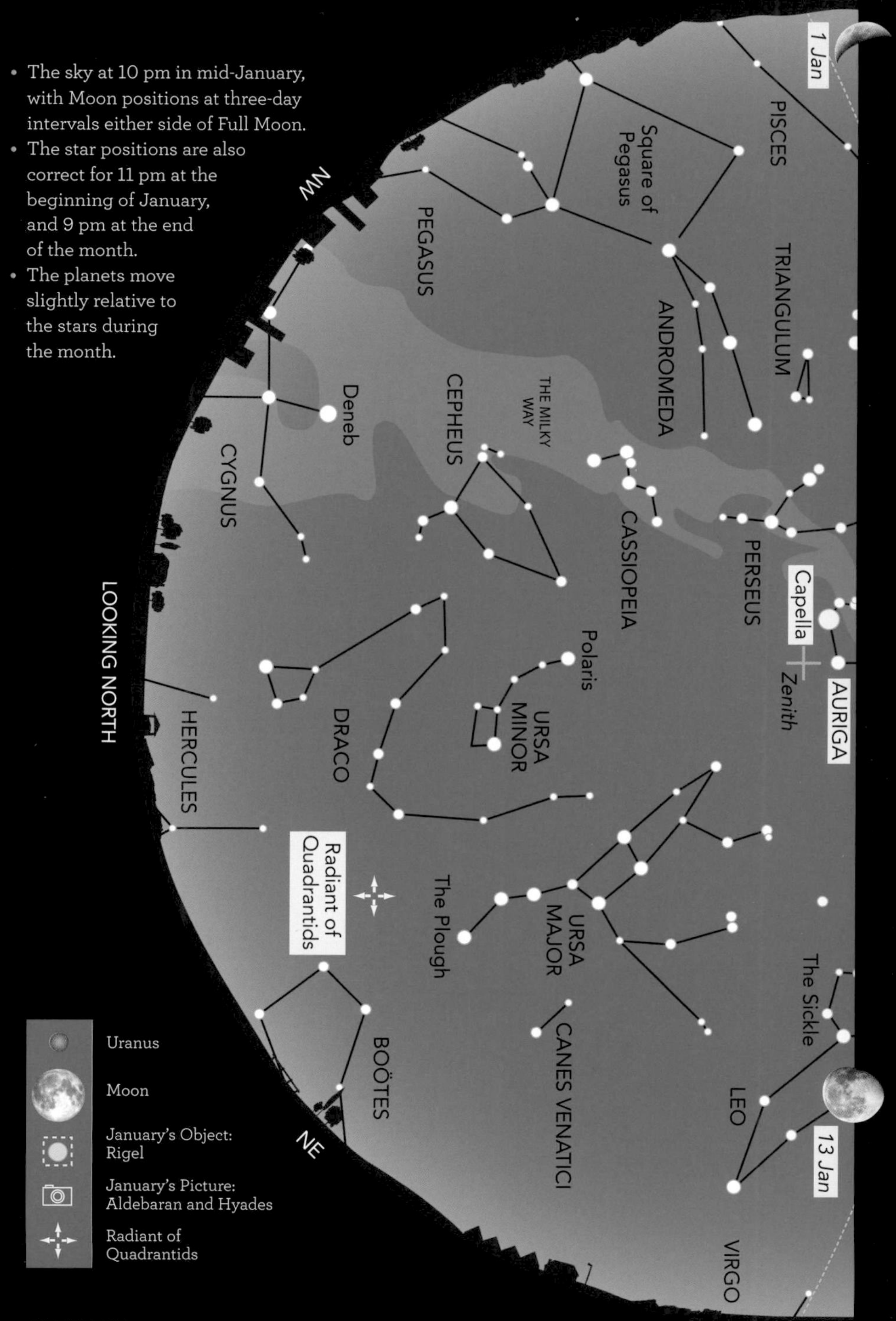

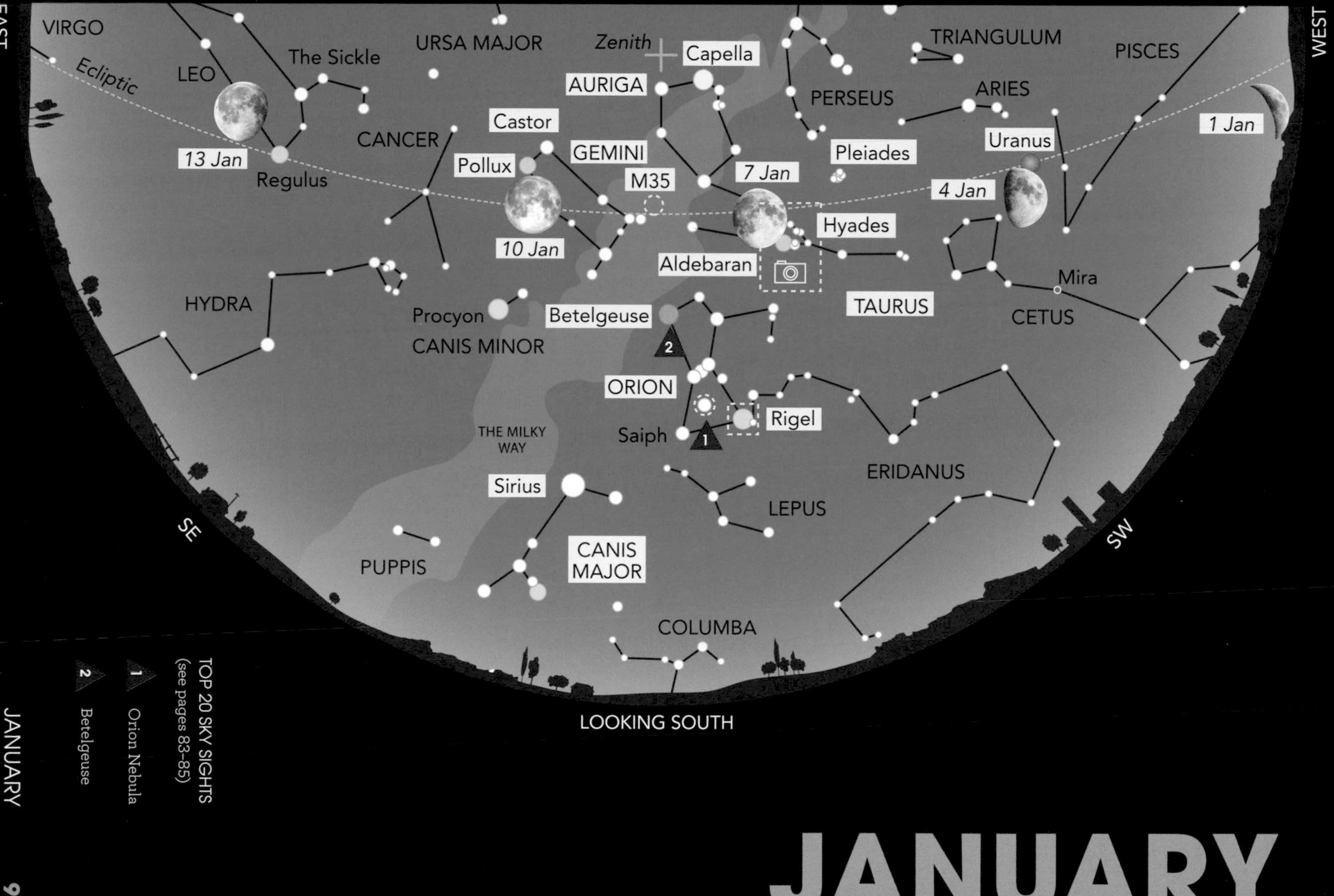

TOP 20 SKY SIGHTS
(see pages 83–85)

1 Orion Nebula

2 Betelgeuse

JANUARY

If ever there was a time to see A-list stars strutting their stuff, it's this month. Heading the cast are **Betelgeuse** and **Rigel** in the hunter **Orion**, with glorious **Sirius** in **Canis Major** (the Great Dog) to the lower left. Forming a giant arc above, you'll find **Aldebaran**, the eye of **Taurus** (the Bull); **Capella**, lighting up **Auriga** (the Charioteer); and **Castor** and **Pollux**, the celestial twins in **Gemini**. They're joined by dazzling Venus – a brilliant lantern in the dusk sky, ushering in the New Year.

JANUARY'S CONSTELLATION

Well known as the Twins, **Gemini** is crowned by the bright stars **Castor** and **Pollux** representing the heads of the celestial brethren, with their bodies running in parallel lines of stars. In legend they were conceived by princess Leda on her wedding night – Castor by her husband, and immortal Pollux by Zeus, who invaded the marital suite disguised as a swan. The devoted pair were placed together for eternity among the stars.

Castor is an amazing family of six stars. Through a small telescope, you can see that it's double. Spectroscopes reveal that both these stars are themselves close pairs. Then there's an outlying star, visible through a telescope, which also turns out to be double.

Pollux – slightly brighter than Castor – boasts a huge planet, Thestias, a mighty world bigger than Jupiter.

The constellation contains a pretty star cluster, **M35**. Even at a distance of nearly 4000 light years, it's visible to the unaided eye, and a fine sight through binoculars or a small telescope.

JANUARY'S OBJECT

Rigel is one of the most iconic stars in the sky. Shining a dazzling blue-white, it marks the bottom corner of Orion's tunic, and it's the most brilliant star in a constellation of a true celestial celebrities. Some 860 light years distant, Rigel is 125,000 times more luminous than our Sun. It is the sixth-brightest star in the heavens.

Venerated in myth and legend all over the world, Rigel's name derives from 'foot' or 'leg' in Arabic. In Central America, Rigel was known as 'the little woodpecker', while in northern Australia, the star was seen as a great ceremonial leader, Unumburrgu, with the other stars of Orion forming his entourage.

Normally around 70 times bigger than the Sun, Rigel pulsates irregularly, due to disturbances in its nuclear core. As a result, its brightness changes by 0.1 of a magnitude.

OBSERVING TIP

If you want to stargaze at this most glorious time of year, dress up warmly! Lots of layers are better than just a heavy coat, as they trap more air close to your skin; while heavy-soled boots with two pairs of socks stop the frost creeping up your legs. It may sound anorak-ish, but a woolly hat prevents a lot of your body heat escaping through the top of your head. And – alas – no hipflask of whisky. Alcohol constricts the veins, and makes you feel even colder.

John Bell caught this image from a dark site in Shropshire with a Sony A7S camera, sporting a Sigma 70–200 mm zoom lens at 200 mm, f/2.8, ISO 5000. He combined five 15-second exposures, because his telescope alignment was slightly off. On this damp night, dew formed on the lens, acting like a diffusion filter to produce the glow around the stars.

Rigel has a companion, Rigel B, lying only 12 light days from the main star. Although it's not particularly faint (magnitude +6.7), this close-in companion is 500 times fainter than Rigel, so you'll need a moderate telescope to spot it. Rigel B is itself double, although its companion is too close to be seen through a telescope, So Rigel is actually a triple-star system.

JANUARY'S TOPIC: AURORAE

Look out for the sky's ultimate light show! The aurora borealis ('northern lights') and its southern equivalent – the aurora australis – are best seen near the Earth's poles. In Aberdeen, these arcs, curtains and rays of green and red light are called the 'Merrie Dancers'. Local folklore puts them down to reflections of sunlight off the polar ice cap.

The cause is the Sun – but in a more fundamental way. Our local star is riddled with magnetic fields, which wind up every 11 years in a serious affliction of dark sunspots. Magnetic loops can short circuit, hurling electrically charged particles at the Earth. Our planet's magnetic field channels them to the poles, where they hit the atmosphere and light up the atoms like gas in a neon tube.

To see the swaying celestial display is an awesome experience (see February's Picture). Though the Sun's magnetic activity is currently near a minimum, it's an unpredictable beast, and you may spot the aurorae if you take a special plane tour or cruise to the Arctic Circle.

Even if you can't make the North Pole, never fear. Powerful aurorae have been reported from the south of France – and we've seen one in Oxfordshire!

JANUARY'S PICTURE

The constellation of **Taurus** boasts two star clusters: the glamorous **Pleiades**, and the less well-known **Hyades**. You can spot the latter easily; its stars lie next to brilliant **Aldebaran** (which has nothing to do with the cluster – it's an orange giant star half as far away).

In this lovely image, John Bell captures the difference in colour between ageing Aldebaran and the youthful Hyades. Its 700 stars are around 625 million years old – ancient for us, but young compared to our venerable Sun (4.6 billion years old).

JANUARY'S CALENDAR

SUNDAY	MONDAY	TUESDAY	WEDNESDAY	THURSDAY	FRIDAY	SATURDAY
			1	2	3 4.45 am First Quarter Moon	4 Quadrantids (am)
5 Earth at perihelion	6	7 Moon near Aldebaran	8	9	10 7.21 pm Full Moon; penumbral lunar eclipse	11 Moon occults Praesepe
12 Moon near Regulus	13 Moon near Regulus	14	15	16 Moon very close to Aldebaran	17 Moon near Spica (am); 12.58 pm Last Quarter Moon	18 Mars near Antares (am)
19	20 Moon near Mars and Antares (am)	21 Moon near Mars and Antares (am)	22	23	24 9.42 pm New Moon	25
26	27 Venus very near Neptune	28 Moon near Venus	29	30	31	

SPECIAL EVENTS

- **Night of 3/4 January:** watch well after midnight, when the Moon has set, for the **Quadrantid meteor shower**. These bright colourful shooting stars are dust particles from the old comet 2003 EH_1, burning up as they impact the Earth's atmosphere.
- **5 January, 7.48 am:** the Earth is closest to the Sun (147 million km away).
- **7 January:** the Moon moves through the Hyades, passing just above Aldebaran (Chart 1a).
- **10 January, around 7 pm:** the lower edge of the Full Moon looks slightly tarnished as it suffers a penumbral eclipse, when some – but not all – of the sunlight falling on the Moon is blocked by the Earth.
- **11 January, about midnight:** the Moon moves over the upper part of the Praesepe star cluster.
- **27 January:** a rare chance to see the brightest and faintest planets together (Chart 1b). Point your binoculars or telescope at Venus and the greenish 'star' very close to the upper right is Neptune (to the lower left is an inverting telescope). The farthest planet lies just 4 arcminutes from Venus, and is 60,000 times dimmer!

JANUARY'S PLANET WATCH

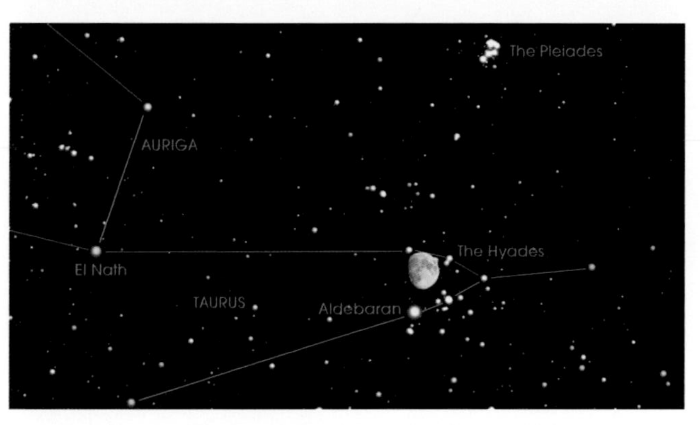

1a *7 January, 6 pm. The Moon lies in the Hyades, close to Aldebaran.*

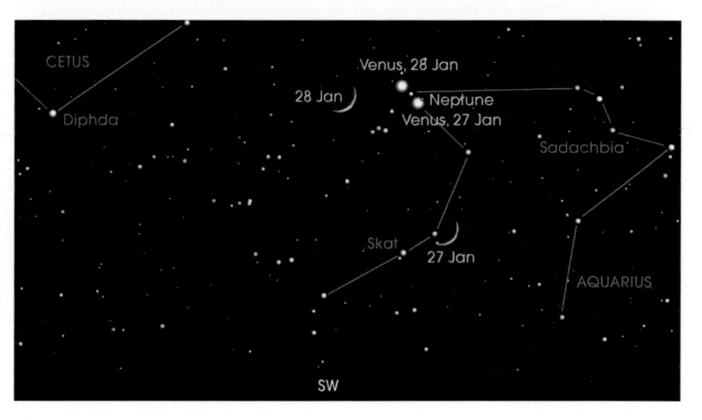

1b *27–28 January, 7 pm. Venus very close to Neptune, with crescent Moon.*

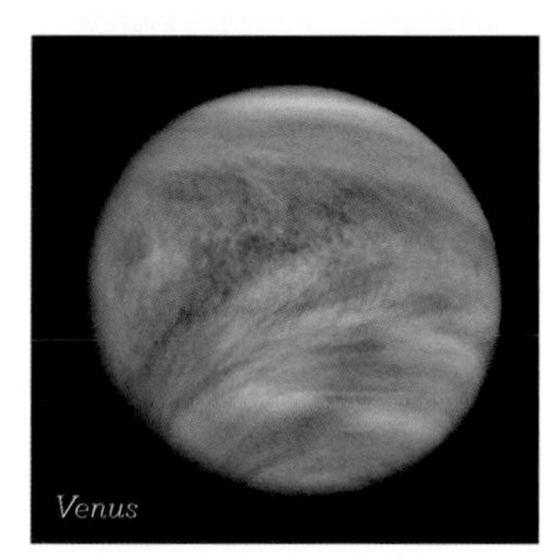

Venus

• **Venus** opens the year shining gloriously in the south-west after sunset. At a brilliant magnitude –4.0, the Evening Star sets at 7 pm at the start of January and 8.30 pm at the end of the month. On 27 January, it passes very close to Neptune (see Special Events).

• On the last few days of January, you may catch **Mercury** near the horizon, to the lower right of Venus. The innermost planet shines at magnitude –1.0, and sets around 5.45 pm.

• **Neptune** (magnitude +7.9) lies in Aquarius, setting about 9 pm. Seek it out on 27 January, when Venus passes very near by (see Special Events).

• Among the faint stars of Pisces, **Uranus** is on the verge of naked-eye visibility at magnitude +5.8, and sets around 1.30 am.

• **Mars** is rising in the south-east at 4.40 am, shining at magnitude +1.4. During January, it tracks from Libra, through the top of Scorpius and into Ophiuchus. Mid-month, the Red Planet passes above the red giant star Antares: its name means 'rival of Mars', so check out (best with binoculars) which carries off the laurels for being the redder!

• **Jupiter** and **Saturn** are too close to the Sun to be seen this month.

- The sky at 10 pm in mid-February, with Moon positions at three-day intervals either side of Full Moon.
- The star positions are also correct for 11 pm at the beginning of February, and 9 pm at the end of the month.
- The planets move slightly relative to the stars during the month.

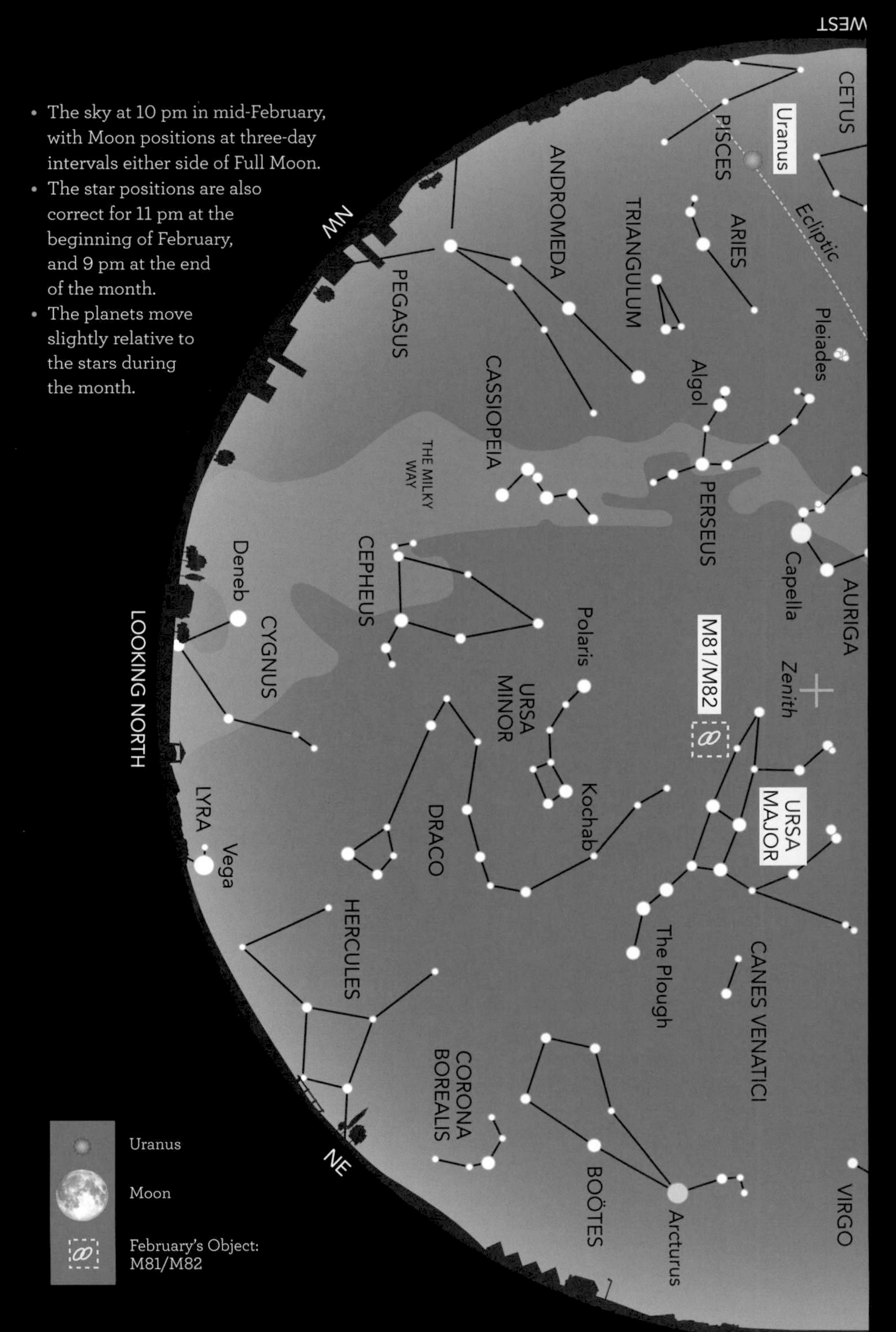

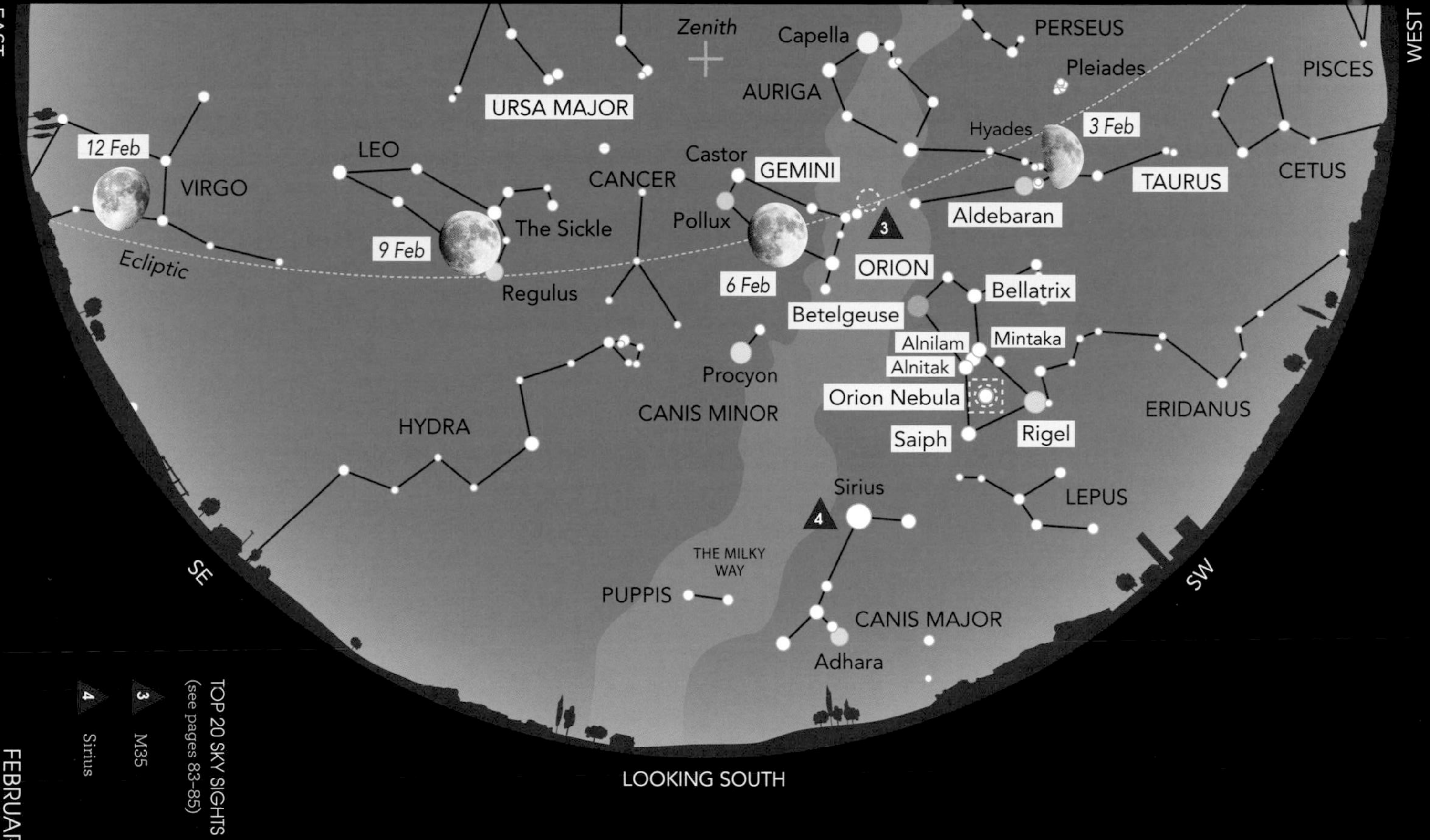

TOP 20 SKY SIGHTS
(see pages 83–85)

3 M35

4 Sirius

Venus is cheering up the dark February nights as a glorious Evening Star, while the winter star patterns – **Orion**, **Taurus** and **Gemini** – are drifting westward. That's a result of Earth's annual orbit around the Sun. Imagine: you're whirling round on a fairground carousel, and looking out around you. At times you spot the ghost train; sometimes you see the roller-coaster; and then you swing past the candyfloss stall. So it is with the sky: as we circle our local star, we get to see different stars and constellations with the changing seasons.

FEBRUARY'S CONSTELLATION

Spectacular **Orion** is a rare star grouping that looks like its namesake – a giant hunter with a sword below his belt, wielding a club. The seven main stars lie in the 'Top 70' brightest stars in the sky, but they're not closely associated – they simply line up, one behind the other.

Closest – at 250 light years – is the fainter of the two stars forming the hunter's shoulders, **Bellatrix**. The other shoulder-star, blood-red **Betelgeuse**, lies 720 light years away. This giant star is a thousand times larger than our Sun, and its fate will be to explode as a supernova.

Slightly brighter, blue-white **Rigel** (Orion's foot) – 860 light years from us – is a young star twice as hot as our Sun and 125,000 times more luminous (see January's Object). **Saiph**, the hunter's other foot, is 650 light years distant.

We must travel around 1300 light years from home to reach the stars of Orion's glittering belt – **Alnitak**, **Alnilam** and **Mintaka** – and the great **Orion Nebula** (see December's Object). The nearest massive 'star factory', the Orion Nebula contains hundreds of embryonic stars, in many cases surrounded by a dusty disc poised to form into a system of planets.

OBSERVING TIP

When you first go out to observe, you may be disappointed at how few stars you can see in the sky. But wait 20 minutes, and you'll be amazed at how your night vision improves. One reason for this 'dark adaption' is that the pupil of your eye grows larger. More importantly, in dark conditions the retina of your eye builds up bigger reserves of rhodopsin, the chemical that responds to light.

FEBRUARY'S OBJECT

A pair of galaxies this month, nuzzling up close and personal to each other in **Ursa Major**. **M81** and **M82** are visible through good binoculars on a really dark night, though you'll need a moderately powerful telescope to reveal them in detail. 'M' stands for Charles Messier, an 18th-century Parisian astronomer who catalogued 103 'fuzzy objects' that misleadingly resembled comets. He did find a handful of comets – but today, he's better remembered for his Messier Catalogue.

M81 is a beautiful, smooth spiral galaxy with curving spiral arms wrapped around a softly glowing core. A small version of the Milky Way, M81 lies 12 million light years away (see March's Picture).

Lying close by is M82 – a spiral galaxy that couldn't be more different. It

Nils Reuther used a Canon EOS 6D Mark I camera, with a Walimex 14 mm f/2.8 lens. He took a 20-second exposure at ISO 800.

looks a total mess, with a huge eruption taking place at its core. Some 300 million years ago, a close encounter with its bigger sibling M81 ripped out streams of gas, and these are now raining back and triggering an explosion of star formation – making M82 the prototype 'starburst galaxy'.

FEBRUARY'S TOPIC: LEAP YEAR

This year, February is blessed with an extra day – as anyone born on 29 February is all too aware! – making 2020 a 'leap year'. We have leap years because there aren't an exact number of days in a year: in fact, a year is 365.2422 days long. Julius Caesar declared that every fourth year should have an extra day (added to poor February, as it's the shortest month). That makes an average of 365.25 days per year – which still isn't quite right. In 1582, Pope Gregory XIII changed the rules, so that a century year is a leap year only if you can divide it by 400. So 2000 was a leap year, while 2100 won't be. Over many centuries, the 'Gregorian calendar year' averages out to 365.2425 days. That's pretty close to the actual length of the year. But, amazingly, back in AD 1079, the Persian astronomer and poet Omar Khayyam devised a calendar – still used in Iran – that averages to 365.2424 days per year, which is more accurate than our Gregorian calendar!

FEBRUARY'S PICTURE

This beautiful aurora – seen from the Isle of Skye – was captured by German visitor Nils Reuther. It shows two glorious green pillars, and green arcs below and above with tantalising hints of red towards the top. There's also a sweep of blue above the pillars.

Aurorae are caused by energetic particles from the Sun lighting up the air above our heads (see January's Topic) – but what creates their beautiful colours? When the solar particles hit the atoms and molecules in the upper atmosphere, they strip electrons away from the nucleus, in a process called 'ionisation'. As the electrons rejoin their parents, they lose energy and emit light.

But not all ionised atoms glow the same colour; it depends on the particular element. Ionised oxygen glows red or green, depending on the altitude; while ionised nitrogen provides blue and purplish tinges.

FEBRUARY'S CALENDAR

SUNDAY	MONDAY	TUESDAY	WEDNESDAY	THURSDAY	FRIDAY	SATURDAY
						1
2 1.41 am First Quarter Moon near the Pleiades	3 Moon near the Pleiades, Hyades, Aldebaran	4	5	6	7	8
9 7.33 am Full Moon near Regulus, supermoon	10 Mercury E elongation	11	12 Moon near Spica	13	14	15 10.17 pm Last Quarter Moon
16	17 Mars between Lagoon and Trifid Nebulae (am)	18 Mars between Lagoon and Trifid Nebulae (am), Moon nearby	19 Moon between Jupiter and Mars (am)	20 Moon near Saturn and Jupiter	21	22
23 3.32 pm New Moon	24	25	26 Moon below Venus	27 Moon near Venus	28 Moon above Venus	29

SPECIAL EVENTS

- **3 February:** to the upper right of the Moon you'll find the Pleiades star cluster, and to its left Aldebaran. The stars of the Hyades are visible between the Moon and Aldebaran. As the Moon sets, at around 3 am, it occults some of the Hyades stars.
- **9 February:** there's a supermoon tonight, with the Full Moon bigger and brighter than usual – but it's not as spectacular as it will be on 8 April, the year's best supermoon.
- **17–18 February:** check out Mars with binoculars or a small telescope, and you'll see the fuzzy patch of the Lagoon Nebula lying half a degree below the Red Planet and the fainter Trifid Nebula the same distance above.
- **18–20 February:** on the morning of 18 February, the crescent Moon lies to the right of Mars; the following morning it's to the left of Mars and the right of Jupiter. Low in the dawn twilight on 20 February, you can catch the thinnest crescent Moon to the lower left of Jupiter, with Saturn to the left of the Moon (Chart 2b).
- **26–28 February:** the crescent Moon appears in the evening sky with Venus, forming a beautiful close pairing with the Evening Star on 27 February.

FEBRUARY'S PLANET WATCH

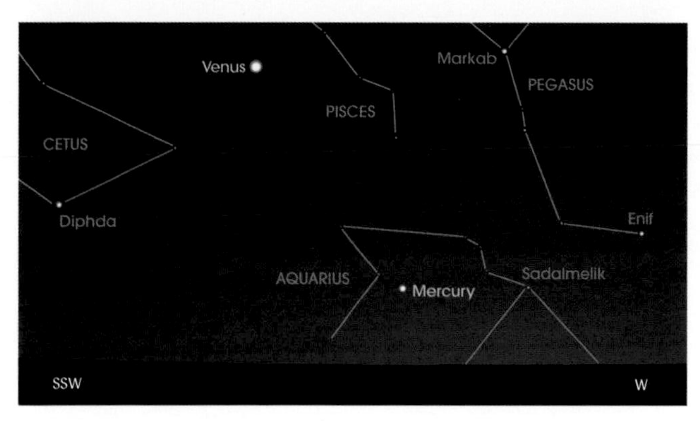

2a 10 February, 6 pm. Mercury at greatest eastern elongation, with Venus.

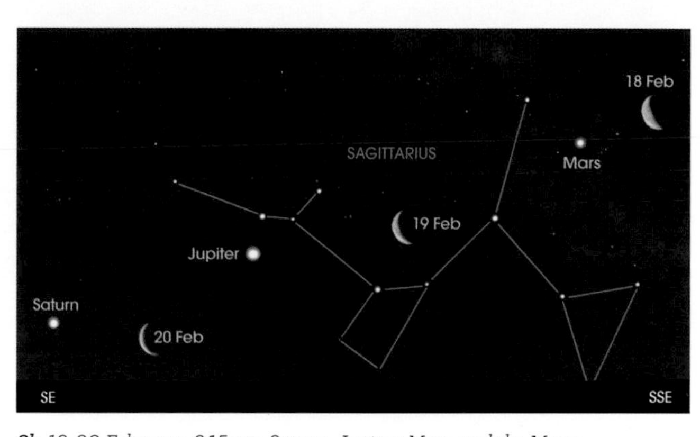

2b 18–20 February, 6.15 am. Saturn, Jupiter, Mars and the Moon.

- **Venus** grows ever brighter as it rises into darker skies after sunset, reaching magnitude –4.2 by the end of February, when the Evening Star sets at 10 pm.
- During the first half of the month, look low down on the horizon to the lower right of Venus, to spot **Mercury** making its best evening appearance of the year. The shy little world sets around 6.50 pm at its greatest elongation from the Sun on 10 February, when it shines at magnitude –0.5 (Chart 2a).
- With binoculars or a telescope, you may just catch **Neptune** (magnitude +7.9) low in the south-west at the start of February, between much brighter Venus and Mercury. Lying in Aquarius, it sets around 7 pm and disappears into the twilight glow by the end of the month.
- **Uranus** lies on the border of Aries and Pisces; at magnitude +5.8, it sets about 11.30 pm.
- There's more planetary action in the morning sky. **Mars** (magnitude +1.2) rises in the south-east about 4.30 am, moving from Ophiuchus into Sagittarius. There's a lovely sight on the mornings of 17 and 18 February, as the Red Planet passes between the Lagoon and Trifid Nebulae (see Special Events).
- Brilliant **Jupiter**, also in Sagittarius and shining at magnitude –1.9, clears the south-eastern horizon around 5.30 am.
- Right at the end of February, you may catch a third planet in Sagittarius, to the lower left of Jupiter: **Saturn** (magnitude +0.6) rising about 6 am.

- The sky at 10 pm in mid-March, with Moon positions at three-day intervals either side of Full Moon.
- The star positions are also correct for 11 pm at the beginning of March, and 10 pm at the end of the month (after BST begins).
- The planets move slightly relative to the stars during the month.

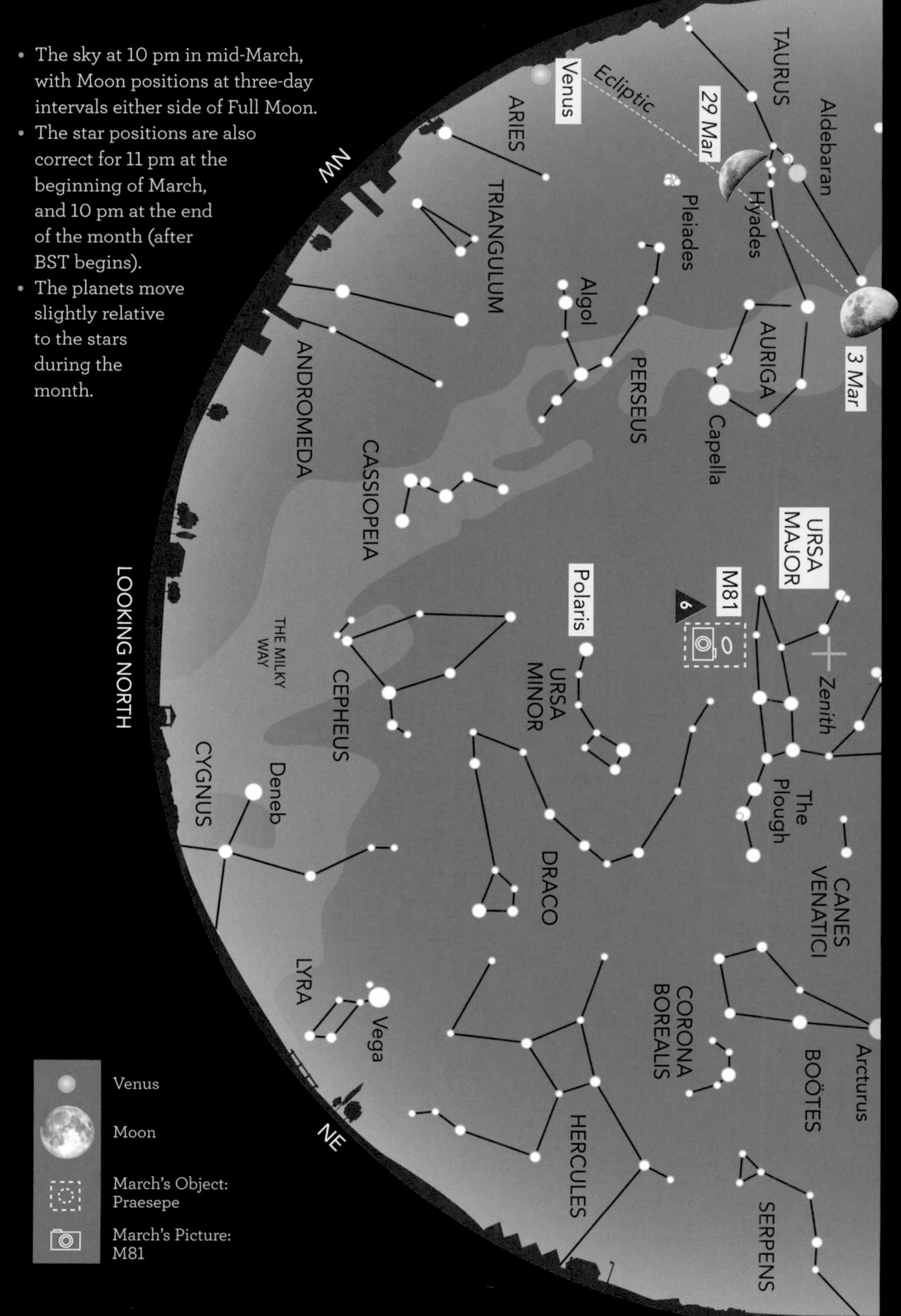

MARCH

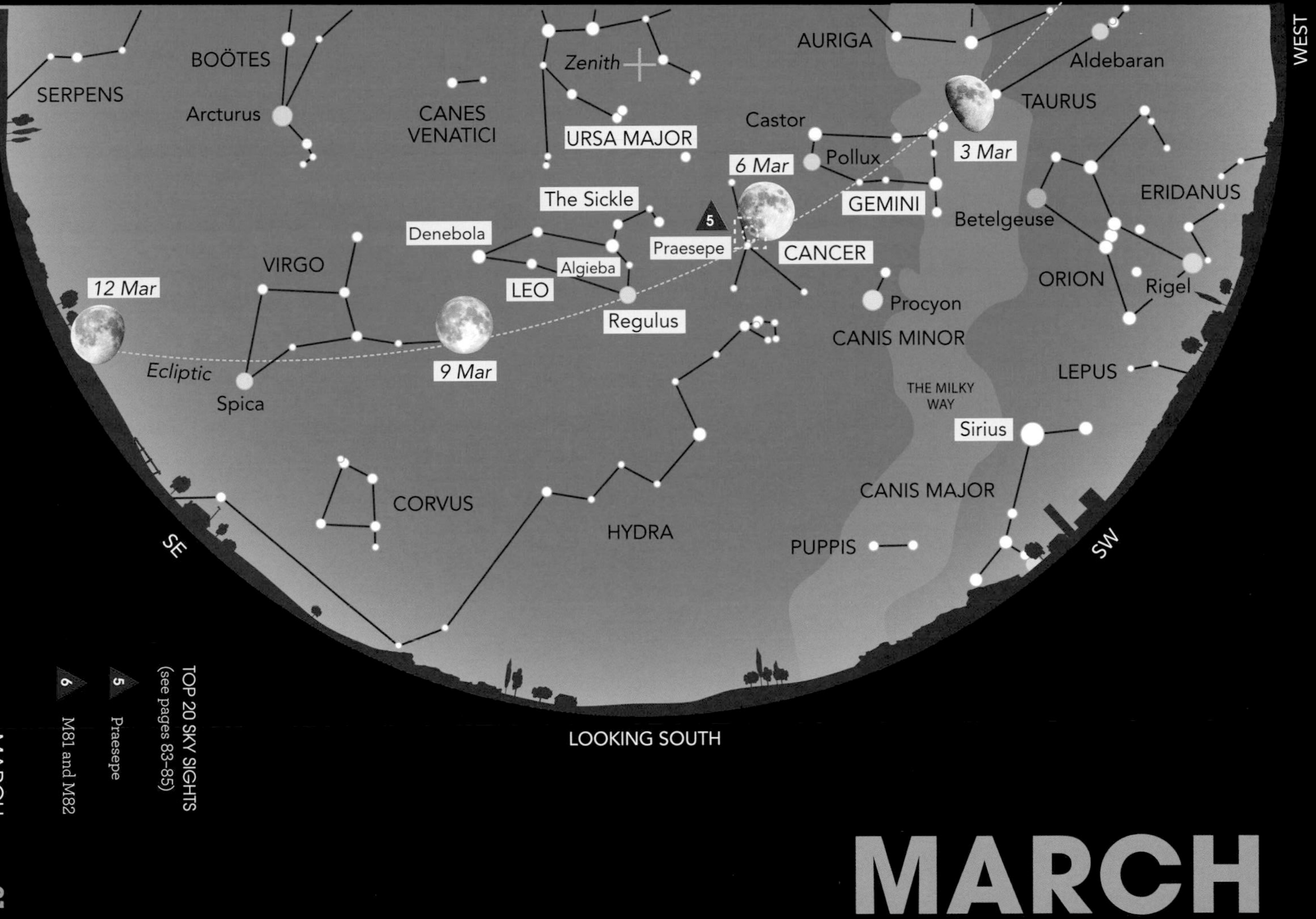

TOP 20 SKY SIGHTS
(see pages 83–85)

5 Praesepe

6 M81 and M82

Spring is here! On 20 March, we celebrate the Equinox, when day becomes longer than night; while British Summer Time starts on 29 March. Though the nights may be shorter, there's still plenty going on in the heavens, from brilliant **Venus** lighting up the western skies after sunset to the trio of Jupiter, Mars and Saturn before dawn.

MARCH'S CONSTELLATION

Like the fabled hunter Orion, **Leo** is one of the rare constellations that resembles the real thing – in this case, an enormous crouching lion. Leo is among the oldest constellations, and commemorates the giant Nemean lion that Hercules slaughtered as the first of his 12 labours. According to legend, the lion's flesh couldn't be pierced by iron, stone or bronze – so Hercules wrestled with the lion and choked it to death.

The lion's heart is marked by the first magnitude star **Regulus**. This celestial whirling dervish spins around in just 16 hours, making its equator bulge remarkably. Rising upwards is **'the Sickle'**, a back-to-front question mark that delineates the front quarters, neck and head of Leo. A small telescope shows that **Algieba**, the star that makes up the lion's shoulder, is actually a beautiful close double star.

The other end of Leo is home to **Denebola**, which means in Arabic 'the lion's tail'. Just underneath the main 'body' of Leo are several spiral galaxies – nearby cities of stars like our own Milky Way. They can't be seen with the unaided eye, but sweep along the lion's tummy with a small telescope to reveal them.

MARCH'S OBJECT

Between **Gemini** and **Leo** lies the faint zodiacal constellation of **Cancer**, the Crab. Concentrate your eyes on the crustacean's centre and, with a little luck, you can see a misty patch. This is **Praesepe** – a dense group of stars whose name literally means 'the manger', but is better known as the Beehive Cluster. If you train binoculars on it, you'll understand why: it really does look like a swarm of bees. On 6 March, the Moon sails in front of the cluster (see Special Events).

Praesepe lies nearly 600 light years away, and contains over 1000 stars, born together some 600 million years ago. Two of its stars have planets in orbit about them, but they are not 'Earths' – instead, they are 'hot Jupiters', gas giants circling close in to their parent star.

Galileo, in 1610, was the first to recognise Praesepe as a star cluster. But the ancient Chinese astronomers obviously knew about it, naming the cluster Zei She Ge – 'the Exhalation of Piled-up Corpses'!

MARCH'S TOPIC: COPERNICUS

Nicolaus Copernicus was the most unlikely person to cause a revolution in the cosmos. But he did – on his dying day in 1543. That's when his book *De Revolutionibus* was published. Copernicus was administrator of Frombork Cathedral in Poland, where he observed the movements of the planets from a tower in the grounds. In his time, the ancient Greek teachings held sway: the Sun and the planets circled the Earth, which was the centre of the Universe.

Copernicus was baffled. His observations didn't agree with the Greek predictions. And – hang on – why was *the Sun* so different from the other moving objects out there? Could it be possible that the planets – including the Earth – circled the Sun instead? He put his theory to the test. Everything fell into place.

But, as a religious man, Copernicus was aware of the Earth's sanctity as being central in the cosmos. Which may be why he left publication of *De Revolutionibus* until he was on his deathbed. Galileo would go on to support Copernicus, and be put under house arrest for his declarations.

A century after Copernicus' death, astronomers all rallied round to agree. The Earth and planets orbit the Sun. His findings ended 1400 years of dogma – and paved the way to how we see the Universe today.

Craig Howman's deep image of M81 was captured on a Sky-Watcher 300PDS telescope, using a Canon 1100D camera at ISO 400. He used a coma corrector and a light pollution filter. This is a 43-exposure image, each lasting 12–19 minutes, giving a total exposure time of 10 hours and 2 minutes.

OBSERVING TIP

This is the ideal time of year to tie down the main compass directions, as seen from your observing site. North is easy – just latch onto Polaris, the Pole Star, using the familiar stars of the Plough (see April's Constellation and Star Chart). And at noon, the Sun is always in the south. But the useful extra in March is that we hit the Spring Equinox, when the Sun rises due east, and sets due west. So remember those positions relative to a tree or house around your horizon.

MARCH'S PICTURE

This beautiful spiral galaxy is **M81**, which nestles among the stars of Ursa Major. Photographer Craig Howman has captured its exquisite delicacy in this image. Paired with galaxy M82 (see February's Object), the duo lie 12 million light years away, and they're part of the M81 group – a small cluster of 34 galaxies similar to our Local Group.

Half the size of the Milky Way, M81 looks like an icon of celestial serenity. But its nucleus harbours a massive black hole – which weighs in at 70 million times the mass of the Sun!

MARCH'S CALENDAR

SUNDAY	MONDAY	TUESDAY	WEDNESDAY	THURSDAY	FRIDAY	SATURDAY
1 Moon near Pleiades, Hyades, Aldebaran	2 7.57 pm First Quarter Moon, near Hyades and Aldebaran	3 Moon occults Crab Nebula	4	5	6 Moon occults Praesepe	7 Moon near Regulus; Venus near Uranus
8 Moon near Regulus	9 5.47 pm Full Moon, supermoon	10	11 Moon near Spica	12	13	14
15 Moon near Antares	16 9.34 am Last Quarter Moon	17	18 Moon near Jupiter and Mars (am)	19 Moon near Saturn (am)	20 Spring Equinox; Mars near Jupiter (am)	21
22	23 Mars very near Pluto (am)	24 9.28 am New Moon; Mercury W elongation; Venus E elongation	25	26	27	28 Moon near Venus and Pleiades
29 BST begins; Moon near Venus, Pleiades, Aldebaran	30	31 Mars near Saturn (am)				

SPECIAL EVENTS

- **1–2 March:** the First Quarter Moon lies near the Pleiades (upper right) and the Hyades with bright Aldebaran.
- **3 March, between 5.20 and 6.20 pm:** the Moon moves in front of the Crab Nebula, though twilight spoils the show.
- **6 March:** the Moon moves across the top of Praesepe between 8 pm and midnight.
- **9 March:** watch out for a supermoon tonight, though it will not be as spectacular as the Full Moon next month.
- **18 March:** there's a lovely sight this morning, as the crescent Moon – rising about 5 am – lies next to brilliant Jupiter, with Mars in between.
- **19 March:** Saturn lies above the crescent Moon in the dawn sky.
- **20 March, 3.49 am:** the Spring Equinox, when day and night are equal.
- **23 March, 5.00 am:** a rare chance to find the distant dwarf planet Pluto when it's almost immediately behind Mars (see Planet Watch).
- **28 March:** a beautiful evening tableau, with the crescent Moon lying next to Venus, with the Pleiades above and Aldebaran and the Hyades to the upper left (Chart 3b).
- **29 March, 1.00 am:** British Summer Time starts – don't forget to put your clocks forward.
- **29 March:** the Moon lies close to Aldebaran, and occults some of the Hyades stars; Venus and the Pleiades lie to the right (Chart 3b).

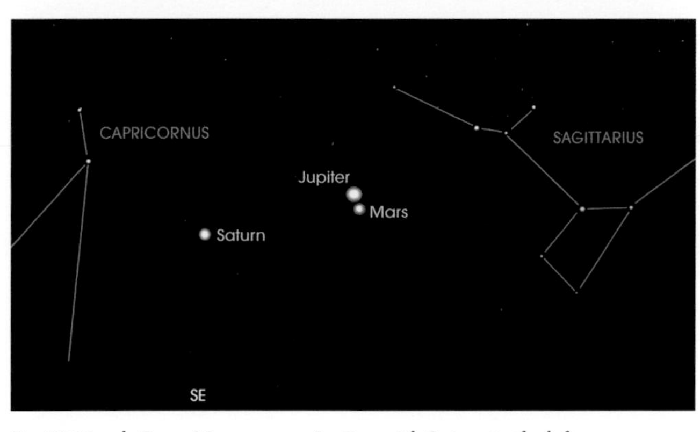

3a 20 March, 5 am. Mars passes Jupiter, with Saturn to the left.

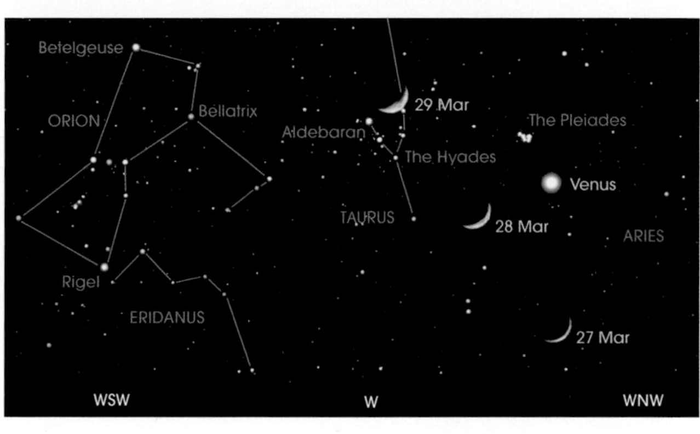

3b 27–29 March, 9 pm. Venus and the Moon, with Aldebaran, the Hyades and the Pleiades.

- **Venus** is brightening in the evening sky, and at greatest elongation on 24 March. At the start of the month it's setting at 10 pm, but by the end of March the Evening Star is visible until after midnight, blazing at magnitude –4.3.
- Lying in Aries, **Uranus** (magnitude +5.9) sets around 9.30 pm, and sinks into the twilight glow by the end of March. Use Venus as a guide on 7 March: swing binoculars to the left of the Evening Star by 2 degrees and Uranus is the greenish 'star'.
- **Jupiter** is lording it over the dawn sky, shining at magnitude –2.0 and rising around 4 am. It's performing a stately dance with **Saturn** and **Mars**, also lying in Sagittarius and appearing above the horizon about the same time. You'll find Saturn to the left of Jupiter, ten times fainter at magnitude +0.7 and shining with a yellowish glow. Mars (magnitude +1.0) starts the month to the right of the others, but the Red Planet speeds leftwards, passing under Jupiter on 20 March (Chart 3a) and Saturn on 31 March.
- On 23 March, use Mars to locate **Pluto**, when it lies just 50 arcseconds away: you'll need a 250 mm telescope (or larger). Between 4.30 and 5.30 am, use a high magnification to observe the Red Planet. The faint star above and to the left, magnitude +14, is the distant dwarf planet (below right in an inverting telescope).
- **Neptune** is lost in the Sun's glare, as is **Mercury** though it's at western elongation on 24 March.

- The sky at 11 pm in mid-April, with Moon positions at three-day intervals either side of Full Moon.
- The star positions are also correct for midnight at the beginning of April, and 10 pm at the end of the month.
- The planets move slightly relative to the stars during the month.

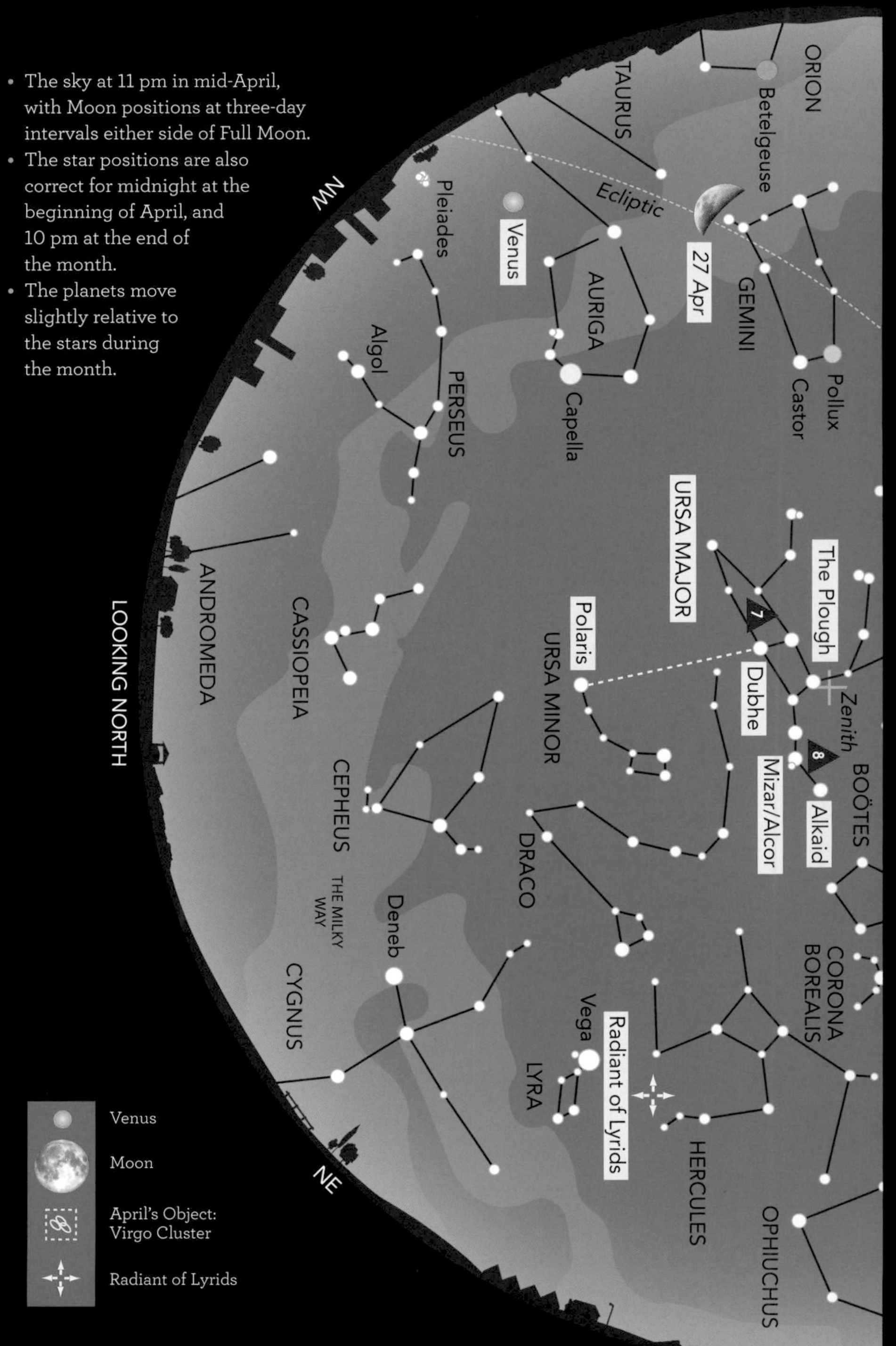

APRIL

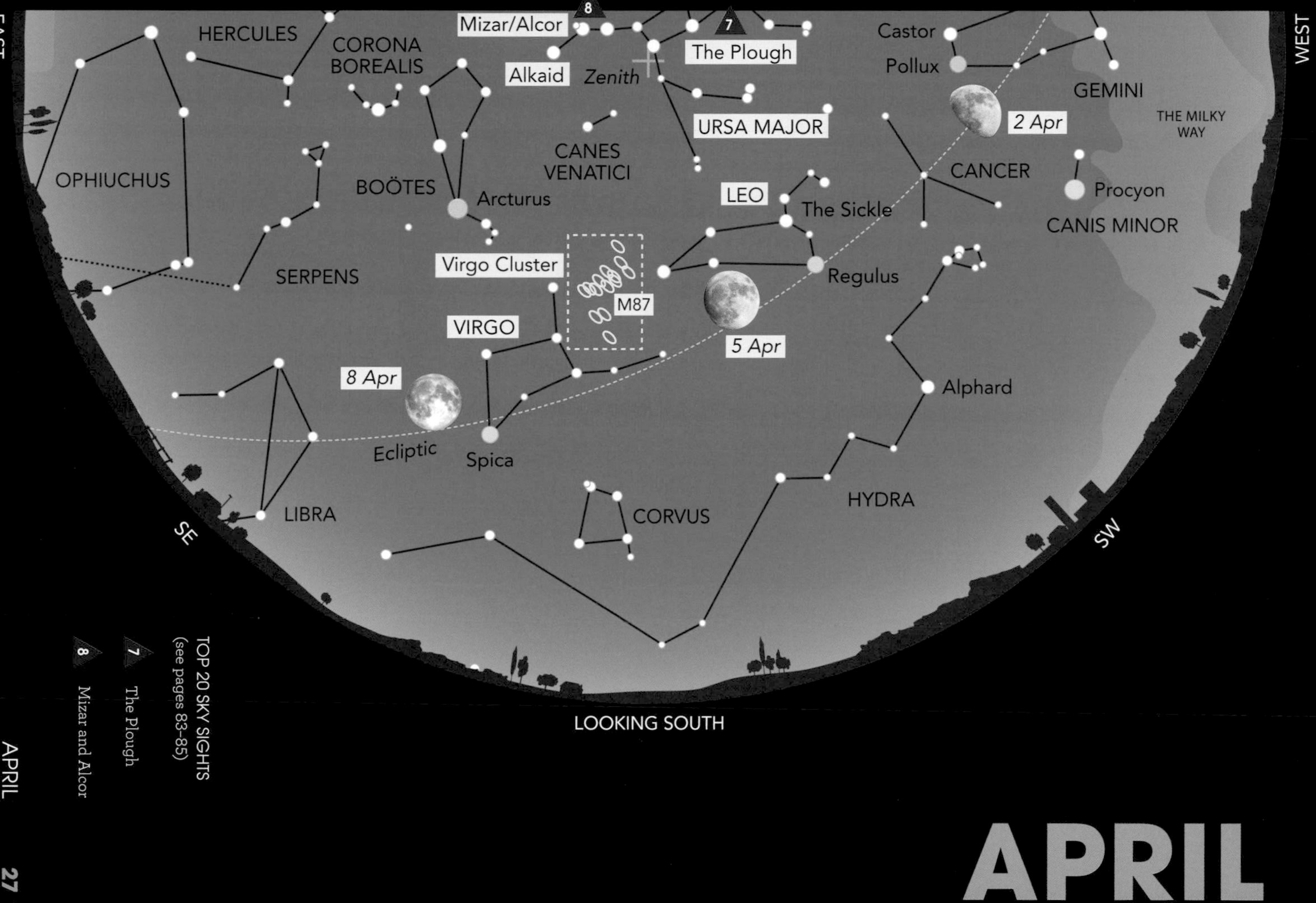

TOP 20 SKY SIGHTS
(see pages 83–85)

- 7 The Plough
- 8 Mizar and Alcor

We're in for a several treats this month, with the brightest Full Moon of the year, plus **Venus** at its most brilliant – and a display of shooting stars. The ancient constellations of **Leo** and **Virgo** dominate the springtime skies. While Leo does indeed look like a recumbent lion, it's hard to envisage Virgo as anything other than a vast 'Y' in the sky!

APRIL'S CONSTELLATION

Ursa Major, the Great Bear, is an internationally favourite constellation. In Britain, its seven brightest stars are called **'the Plough'**. Children today generally haven't seen an old-fashioned horse-drawn plough, and we've found them naming this star pattern 'the saucepan'. In North America, it's known as 'the Big Dipper'.

The Plough is the first star pattern that most people get to know. It's always on view in the northern hemisphere, and the two end stars of the 'bowl' of the Plough point directly towards the Pole Star, **Polaris**, which always lies due north.

Ursa Major is unusual in a couple of ways. First, it contains a double star that you can split with the naked eye: **Mizar**, the star in the middle of the bear's tail (or the handle of the saucepan) has a fainter companion, **Alcor**.

OBSERVING TIP

Venus is at its glorious best this month. Through a small telescope, you'll be able to make out its crescent shape. But don't wait for the sky to get totally dark. Seen against a black sky, the cloud-wreathed world is so brilliant it's difficult to make out any details. You're best off viewing Venus soon after the Sun has set, when the Evening Star first becomes visible in the twilight glow. Through a telescope, the planet then appears less dazzling against a pale blue sky.

And – unlike most constellations – the majority of the stars in the Plough lie at the same distance and were born together. Leaving aside **Dubhe** and **Alkaid**, the others are all moving in the same direction (along with brilliant Sirius, which is also a member of the group). Over thousands of years, the shape of the Plough will gradually change, as Dubhe and Alkaid go off on their own paths.

APRIL'S OBJECT

If you have a small telescope, sweep the 'bowl' formed by **Virgo**'s 'Y' shape, and you'll detect dozens of fuzzy blobs. These are just a handful of the galaxies making up the **Virgo Cluster**: our closest giant cluster of galaxies, some 54 million light years away.

Galaxies are gregarious. Thanks to gravity, they like living in groups. Our Milky Way inhabits a small cluster of about 50 mainly dwarf galaxies called the Local Group. But the Virgo Cluster is in a different league: it's like a vast galactic swarm of bees. What's more, its powerful gravity holds sway over the smaller groups around – including our Local Group – making a cluster of clusters of galaxies, the Virgo Supercluster.

The 2000 galaxies in the Virgo Cluster are also mega. Many of them are spirals like our Milky Way, but some are more spectacular – including the Sombrero, which looks just like a hat. The

Tom Heaton used a Canon 5D MKIII camera, with a Sigma 10–20 mm lens. With ISO 800 at f/3.5, he gave the fireball a 30-second exposure.

heavyweight is M87, a giant elliptical galaxy that emits a 5000-light-year long jet of gas travelling at one-tenth the speed of light.

APRIL'S TOPIC: 'DARK SIDE OF THE MOON'

There ISN'T a Dark Side of the Moon! The misconception is based on the fact that – as a result of gravitational braking – we only see one side of the Moon (the one bearing 'the face of the Man in the Moon'). But as the Moon circles the Earth, all of its surface catches the Sun over the course of a month.

Our first view of the 'farside' came in 1959, when Russia's Luna 3 took a peek. It revealed a densely cratered surface with few of the huge lava-filled basins that we see on the 'nearside'.

The best description of the farside came from Lunar Module pilot Bill Anders, when he swung his Apollo 8 spacecraft around the Moon in 1968. 'The back side looks like a sand pile my kids have played in . . . It's all beat up, no definition, just a lot of bumps and holes.'

APRIL'S PICTURE

This stunning fireball – crossing the Milky Way above a loch bathed in twilight – was captured by international landscape photographer Thomas Heaton on a trip to Dumfries and Galloway with his girlfriend Charlotte, on 21 September 2012.

Fireballs are larger cousins of the shooting stars you can see any night. As these pebble-sized chunks of iron or rock smash into atmosphere, they blaze a brilliant trail, which can outshine the Full Moon.

Tom reports: 'Charlotte saw it first. She screamed out: "Thomas! Look! Look!"' They thought at first it was a crashing plane.

'The fireball was red, blue and green in colour,' he continues, 'with dazzling sparks flying off it. CAMERA! I sprinted across the car park, grappled with my tripod, and hit the shutter.' He had no time to change the settings, but luckily his camera had been set up to image the Milky Way.

'It was a long 30 seconds, but when I viewed my screen, I knew I had captured a very rare and special event.'

APRIL'S CALENDAR

SUNDAY	MONDAY	TUESDAY	WEDNESDAY	THURSDAY	FRIDAY	SATURDAY
			1 11.21 am First Quarter Moon	2	3 Venus amid Pleiades	4 Moon near Regulus
5	6	7 Moon near Spica	8 3.35 am Full Moon, supermoon, near Spica	9	10	11
12	13	14 11.56 pm Last Quarter Moon	15 Moon near Jupiter and Saturn (am)	16 Moon near Mars (am)	17	18
19	20	21 Lyrids	22 Lyrids (am)	23 3.26 am New Moon	24	25 Moon between Pleiades and Aldebaran
26 Moon near Venus	27	28 Venus at maximum brightness	29	30 9.38 pm First Quarter Moon		

SPECIAL EVENTS

- **3 April:** we are treated to a rare and beautiful sight – Venus in front of the Pleiades (see Planet Watch).
- **8 April:** the brightest Moon of 2020. This year there are four supermoons – Full Moons within 367,607 km of the Earth – between February and May, but this month's is the closest, with the Moon just 357,029 km away. It's some 30 per cent brighter than the faintest Full Moon.
- **15 April, in the morning sky:** Jupiter lies just above the crescent Moon to the right, with Saturn to the Moon's upper left (Chart 4b).
- **16 April:** Mars is above the crescent Moon in the morning sky (Chart 4b).
- **Night of 21/22 April:** maximum of the **Lyrid meteor shower**, which appears to emanate from the constellation Lyra. It's a good year for observing these meteors – dusty debris from Comet Thatcher burning up in the Earth's atmosphere – as moonlight doesn't interfere.
- **25 April, just after sunset:** you'll find the thinnest crescent Moon between the Pleiades (right) and Aldebaran and the Hyades (left), with Venus above.
- **26 April:** there's a gorgeous pairing of Venus and the crescent Moon, in the west after sunset.
- **This month,** the European spacecraft BepiColombo swings past the Earth en route to Mercury.

APRIL'S PLANET WATCH

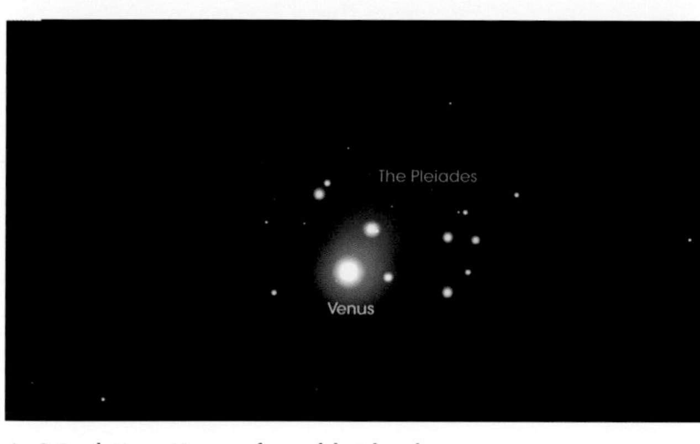

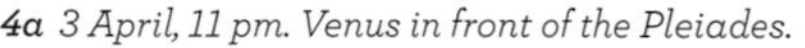
4a *3 April, 11 pm. Venus in front of the Pleiades.*

AQUARIUS
CAPRICORNUS
SAGITTARIUS
Saturn
Jupiter
Mars
14 Apr
15 Apr
16 Apr
SE
SSE

4b *14–16 April, 4.30 am. Mars, Saturn, Jupiter and the Moon.*

• **Venus** is at its most spectacular this month. The Evening Star reaches its maximum brilliance of magnitude –4.5 on 28 April, and stays above the horizon until 0.30 am. Through a small telescope, you'll see our neighbour world grow steadily larger as it approaches the Earth, while its shape changes from half-lit to crescent.

• On 3 April, Venus passes in front of the Pleiades (Chart 4a). The Evening Star is so bright that to the unaided eye it may just look as though Venus has gone fuzzy; but binoculars or a small telescope will reveal the glory of the brilliant planet accompanied by a swarm of stars.

• The other three planets visible this month are all rising just before the Sun.

• Brightest is **Jupiter**, shining at magnitude –2.2 in Sagittarius: the giant planet rises about 3 am. Half an hour later, **Saturn** (magnitude +0.6) rises above the horizon in Capricornus.

• At the beginning of April, **Mars** – also in Capricornus – lies just below Saturn and rises at the same time. It's then slightly fainter (magnitude +0.8) than the ringworld. During the month the Red Planet scurries towards the left, growing brighter to finish April a little more brilliant than Saturn at magnitude +0.4.

• **Mercury**, **Uranus** and **Neptune** lie too close to the Sun to be visible.

- The sky at 11 pm in mid-May, with Moon positions at three-day intervals either side of Full Moon.
- The star positions are also correct for midnight at the beginning of May, and 10 pm at the end of the month.
- The planets move slightly relative to the stars during the month.

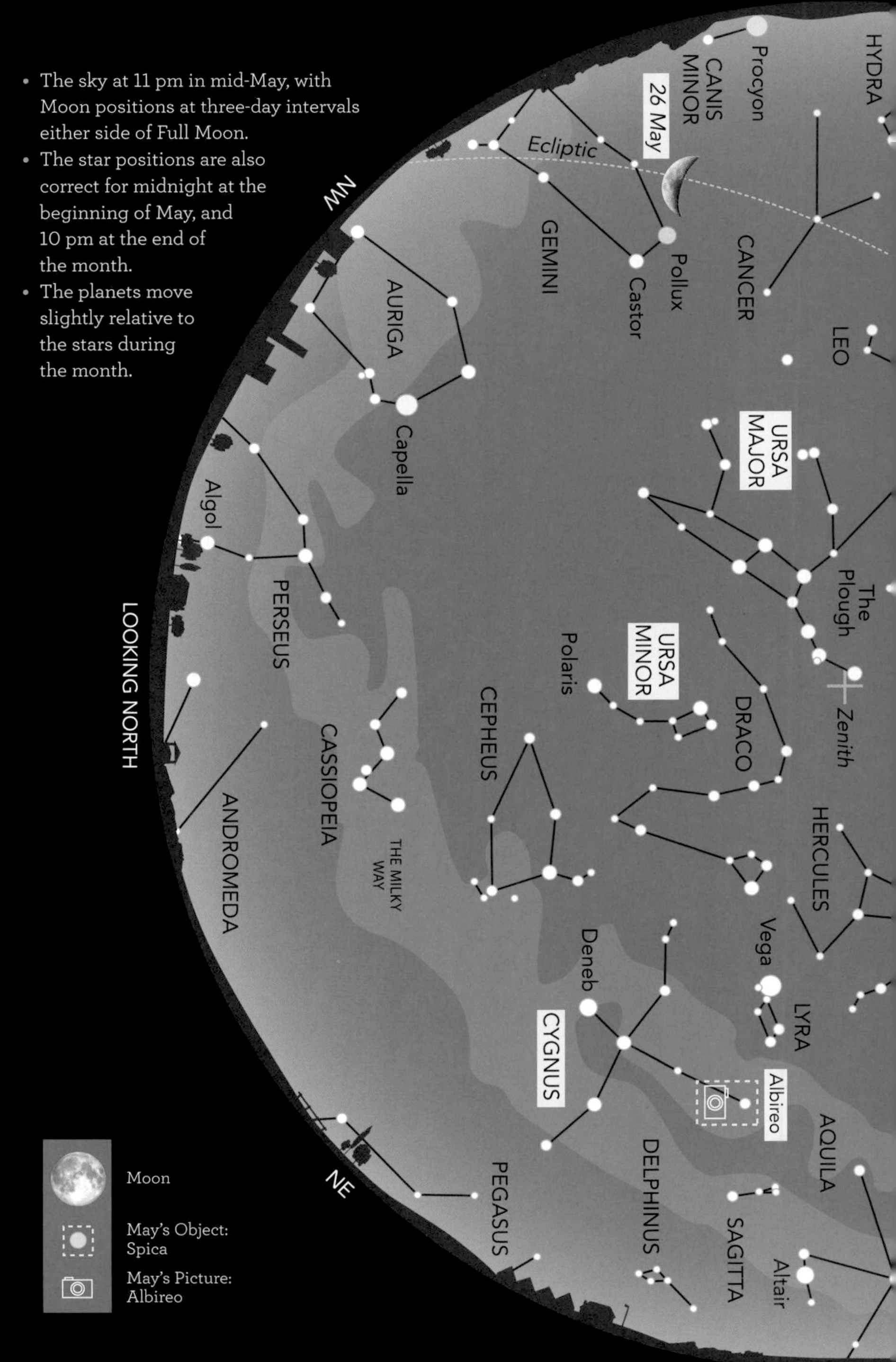

MAY

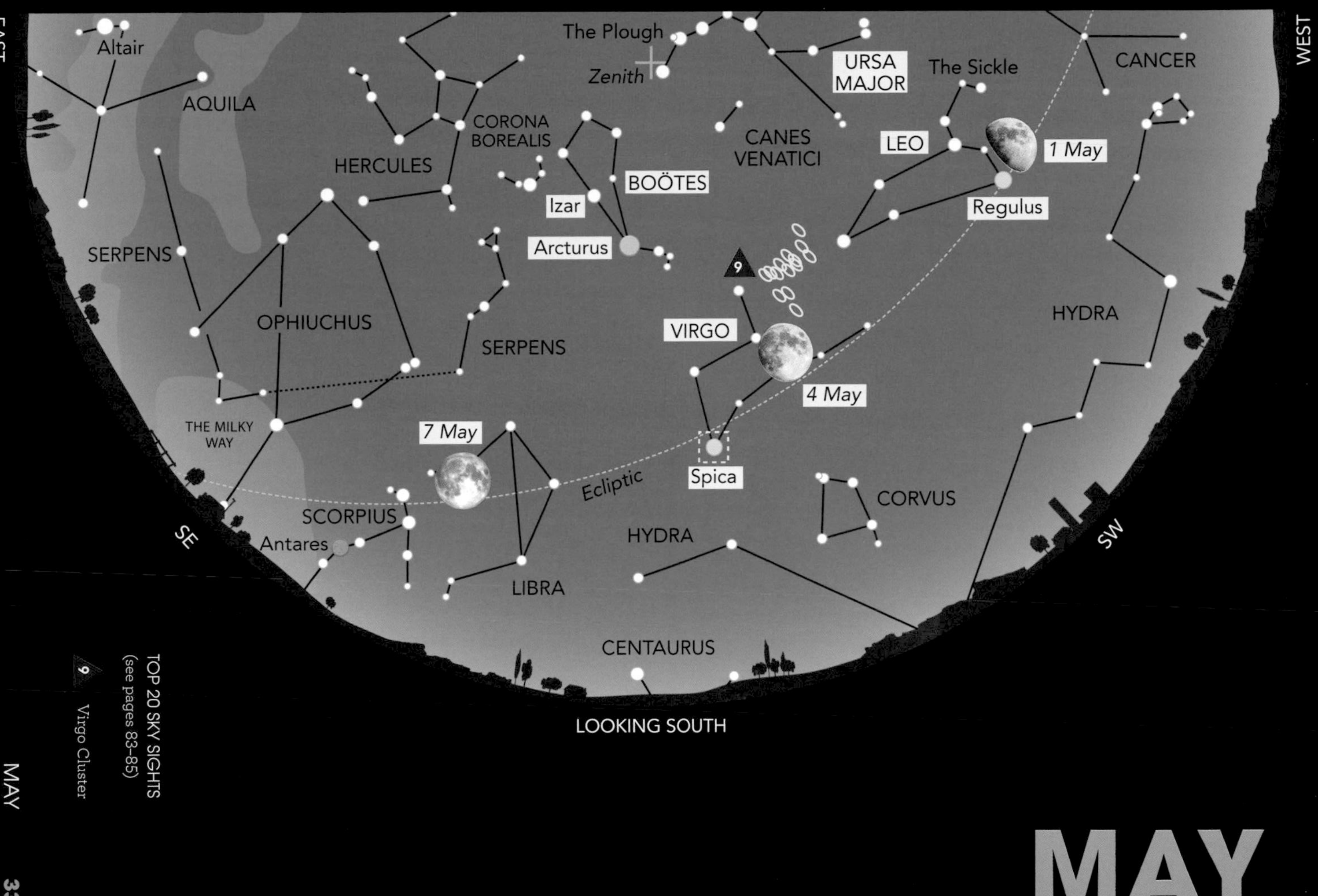

TOP 20 SKY SIGHTS
(see pages 83–85)

9 Virgo Cluster

We have not just one, but two evening stars this month, as brilliant Venus is joined by the elusive little planet Mercury. Star-wise, orange-coloured **Arcturus**, the principal star of **Boötes**, forms a giant triangle with two blue-white stars: **Spica**, in **Virgo**, and **Leo**'s shining light **Regulus**.

MAY'S CONSTELLATION

Boötes, the Herdsman, is shaped rather like a kite. It was mentioned in Homer's *Odyssey*, and its name refers to the fact that Boötes seems to 'herd' the stars that lie in the northern part of the sky.

The name of the brightest star, **Arcturus**, means 'bear driver'. It apparently 'drives' the two bears (**Ursa Major** and **Ursa Minor**) around the sky as the Earth rotates. Arcturus is the fourth brightest star in the whole sky, and it's the most brilliant star you can see on May evenings. A red giant star in its old age, Arcturus lies 37 light years from us, and shines 170 times more brilliantly than the Sun.

The star at the ten o'clock position from Arcturus is called **Izar**, meaning 'the belt'. Through a good telescope, it appears as a gorgeous double star – one star yellow and the other blue.

MAY'S OBJECT

The constellation of **Virgo** is depicted as a virtuous maiden holding an ear of corn which is marked by the brilliant blue-white star **Spica**. Her body rises to the upper right of Spica as a line of stars that splits halfway along, to create a great cosmic 'Y'.

Spica is a pretty exceptional star. Ten times heavier than the Sun, it's also more than 12,000 times more luminous. And its surface gases rage at 25,000°C (compared to the Sun's more modest 5,500°C).

Jamie Cooper captured this image at the prime focus of an 8 inch (200 mm) f/5 Sky-Watcher Newtonian telescope. He used a Canon 6D camera, which he exposed for 20 seconds at ISO 2000. Jamie then stacked and averaged ten images to reduce image noise, and finished it off in Photoshop.

Just 250 light years away, Spica is close on the celestial scale – so it's easy to study. Astronomers have discovered that the star is double, but with a difference. The two stars in the system are so close that they circle one another in just four days. The result? Gravity distorts the stars into egg shapes!

MAY'S TOPIC: ANALYSING THE UNIVERSE

People often ask us: 'How can you know so much about the Universe when it's so far away?' Philosophers – like the 19th-century Auguste Comte – thought that a full understanding of the stars was beyond us. But even as he made this claim, a young German glassmaker had evidence to prove him wrong.

Joseph von Fraunhofer made exquisitely accurate prisms, which he tested on the Sun. Prisms spread light out into a rainbow – the spectrum – ranging from long-wavelength red rays to short-wavelength blue. Fraunhofer was dismayed when he saw his rainbows crossed by dark, vertical lines. Was his craftsmanship at fault? Not being a scientist, he mapped the positions of 570 lines, and went back to glassmaking.

Half a century later, scientists Robert Bunsen and Gustav Kirchhoff at the University of Heidelberg were intrigued by Fraunhofer's discovery. They heated up various elements – like calcium, sodium and potassium – in the flame of the newly invented Bunsen burner and then passed the elements' light through a prism.

Each emitted a pattern of lines – unique to the element. And the scientists realised that their lines matched those that Fraunhofer had discovered in the Sun's spectrum.

OBSERVING TIP

It's always fun to search out the 'faint fuzzies' in the sky – galaxies, star clusters and nebulae. But don't even think of observing dark sky objects around the time of Full Moon – its light will drown them out. You'll have the best views near New Moon: a period astronomers call 'dark of Moon'. When the Moon is bright, though, there's still plenty to see: focus on planets, bright double stars – and, of course, the Moon itself. Check our month-by-month Calendars for the Moon's phases.

At last: astronomers had a way to 'fingerprint' the stars. Not only is a star's spectrum a guide to its makeup; it also reveals the star's age, temperature and how fast it's moving. Spectroscopy is key to understanding our cosmos, and the lines the young glassmaker discovered are called 'Fraunhofer Lines' in his honour.

MAY'S PICTURE

Albireo is the most beautiful double star in the sky. It marks the head of the Swan in the constellation **Cygnus**, and a small telescope easily reveals its two components. Albireo lies 415 light years away, and although the jury is still out as to whether the two stars actually orbit one another, or merely lined up by chance, they make a stunning sight.

A bright golden-yellow star nuzzles up to a fainter, sapphire-blue one. The yellow star is 70 times heavier than the Sun, and 1300 times brighter. Some 230 times more luminous than the Sun, the blue star boasts a temperature of 13,000°C.

MAY'S CALENDAR

SUNDAY	MONDAY	TUESDAY	WEDNESDAY	THURSDAY	FRIDAY	SATURDAY
31					1 Moon near Regulus	2
3	4	5 Moon near Spica	6 Eta Aquarids (am)	7 11.45 am Full Moon, supermoon	8	9
10	11	12 Moon near Jupiter and Saturn (am)	13 Moon near Jupiter and Saturn (am)	14 3.02 pm Last Quarter Moon	15 Moon near Mars (am)	16
17	18	19	20	21 Mercury near Venus	22 6.39 pm New Moon; Mercury near Venus	23
24 Moon near Venus and Mercury	25	26 Moon near Castor and Pollux	27 Moon near Praesepe	28	29 Moon near Regulus	30 4.30 am First Quarter Moon

SPECIAL EVENTS

- **Night of 5/6 May:** shooting stars from the Eta Aquarid meteor shower – tiny pieces of Halley's Comet burning up in Earth's atmosphere – fly across the sky in the early hours of the morning. This year, bright moonlight spoils the show.
- **7 May:** there's a supermoon tonight, though the Full Moon is not as big and bright as it was on 8 April.
- **12–13 May:** the brilliant 'star' near the Moon in the morning sky is giant planet Jupiter. Saturn lies to its left, some 15 times fainter (Chart 5a).
- **15 May, before dawn:** you'll find Mars above the crescent Moon (Chart 5a).
- **21–22 May:** look near brilliant Venus to spot a second 'evening star' – the 30-times-fainter Mercury. The innermost planet lies below Venus on 21 May and to its left on 22 May (Chart 5b).
- **24 May:** as it grows dark, check out the north-western horizon to enjoy the gorgeous sight of the narrow crescent Moon with glorious Venus to its right and the dimmer Mercury between them (Chart 5b).
- **27 May:** the Moon grazes the top edge of Praesepe.

MAY'S PLANET WATCH

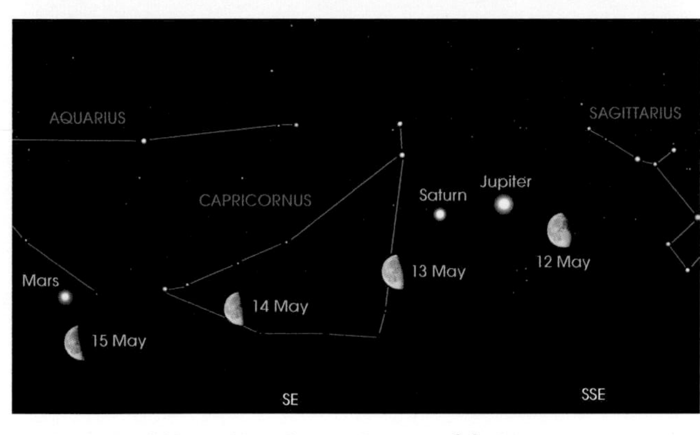

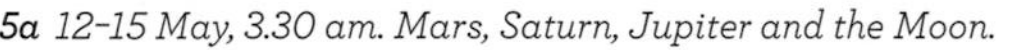
5a 12–15 May, 3.30 am. Mars, Saturn, Jupiter and the Moon.

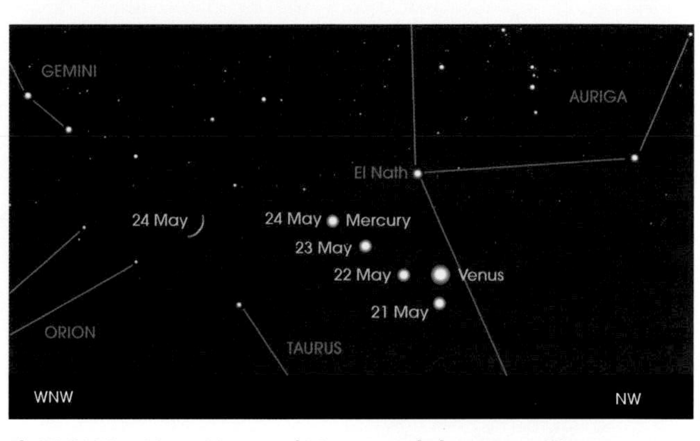

5b 21–24 May, 10 pm. Venus and Mercury, with the crescent Moon.

• Brilliant **Venus** (magnitude –4.5) starts the month as an Evening Star, shining all evening long and setting after midnight. Through a telescope or good binoculars you can make out its crescent shape. But Venus quickly swings towards the Sun, and disappears by the end of May.

• In the second half of May, **Mercury** joins Venus in the evening sky. Around 15 May you can spot Mercury to the lower right of Venus, setting about 10 pm. The innermost planet moves rapidly upwards, passing to the left of Venus on 21 and 22 May when it's at magnitude –0.6. Mercury continues to rise higher and fade; by the end of the month it's magnitude +0.2 and setting at 11 pm.

• The giants of the Solar System, **Jupiter** and **Saturn** rise together at around 1 am. Jupiter, the brighter at magnitude –2.5, lies in Sagittarius while Saturn (magnitude +0.5) inhabits the neighbouring constellation Capricornus.

• **Mars** starts the month near Saturn, also in Capricornus, but heads leftwards during May to end up in Aquarius. Shining at magnitude +0.2, the Red Planet can be seen rising just before 3 am.

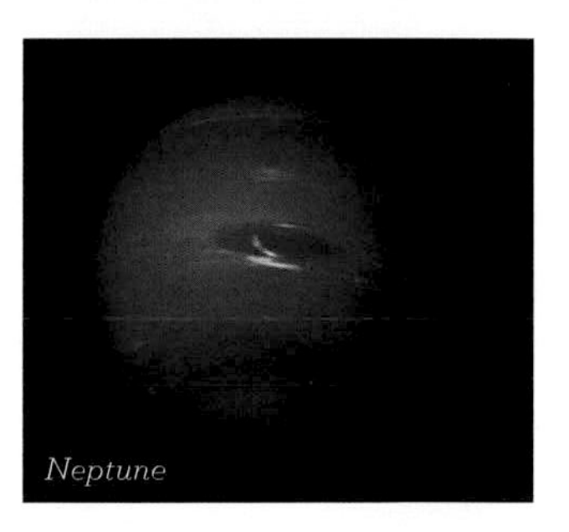

Neptune

• At magnitude +7.9, **Neptune** lies in Aquarius, rising around 3 am. **Uranus** is lost in the Sun's glare this month.

- The sky at 11 pm in mid-June, with Moon positions at three-day intervals either side of Full Moon.
- The star positions are also correct for midnight at the beginning of June, and 10 pm at the end of the month.
- The planets move slightly relative to the stars during the month.

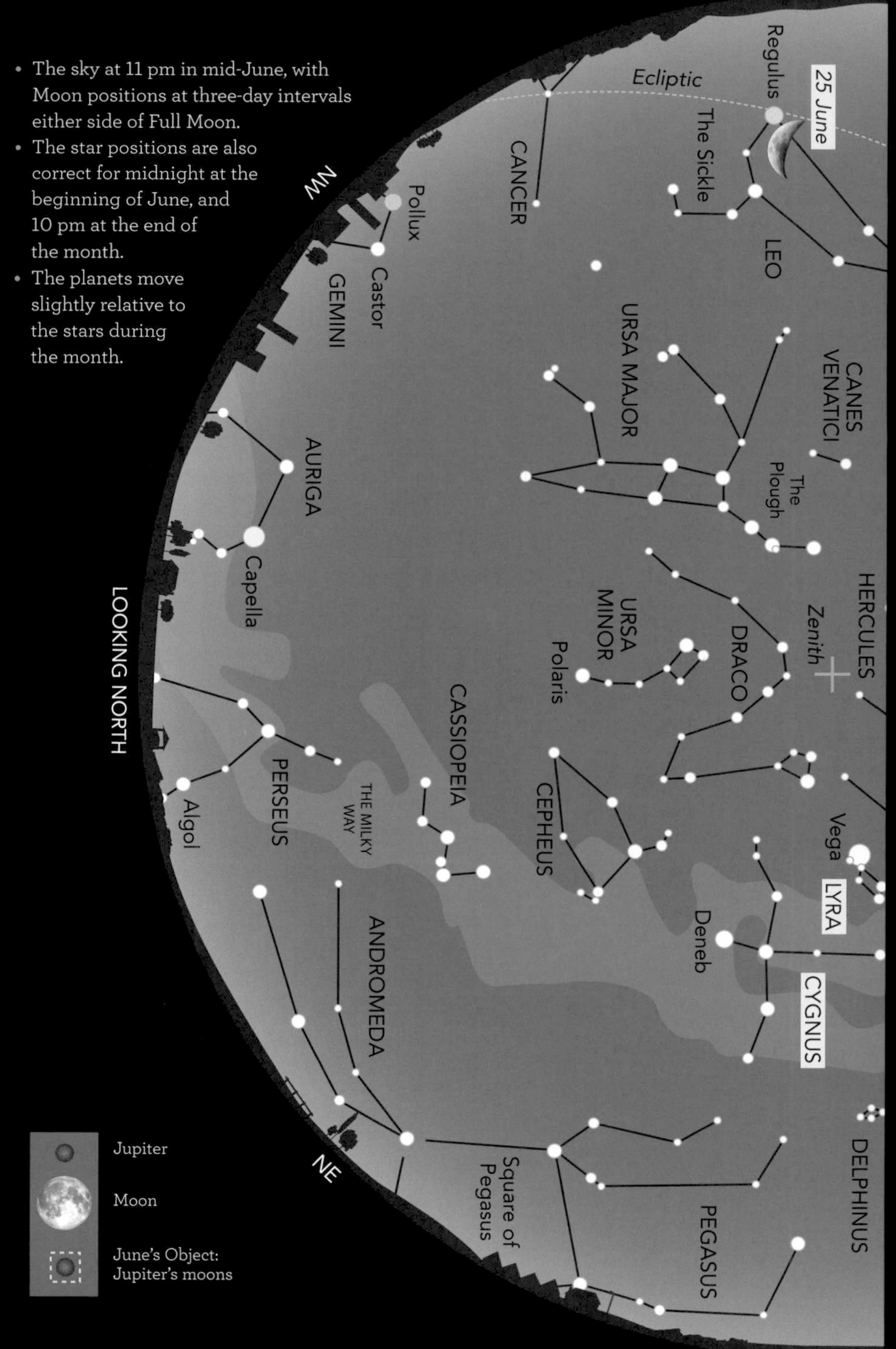

TOP 20 SKY SIGHTS
(see pages 83–85)

10 Antares

11 M13

With the sky never quite getting dark – especially in the north of the country – it's not the greatest month for spotting faint stars. But take advantage of the soft, warm weather to acquaint yourself with the lovely summer constellations of **Hercules**, **Scorpius**, **Lyra**, **Cygnus** and **Aquila**. And if you're up late, you'll be treated to the giant planets **Jupiter** and Saturn rising in the south-east.

JUNE'S CONSTELLATION

For one of antiquity's superheroes – famous for his 12 heroic labours – the celestial version of **Hercules** is a wimp. While Orion is all strutting masculinity in the winter sky, summer's Hercules is a poor reflection – and upside-down to boot.

Dig a little deeper, however, and you'll find a fascinating constellation. Below the hero's rectangular 'body' lies **Rasalgethi**, marking Hercules' head. About 400 times the Sun's girth, this giant star is flopping and billowing in its death throes, varying in brightness from third to fourth magnitude over a period of 90 days. It is circled by a companion star, which takes 3600 years to complete an orbit. There's evidence for a total of four stars in the system.

The jewel in Hercules's crown, though, is **M13** – a bee-like swarm of a third of a million red giant stars, some of the oldest denizens of our Galaxy. One of the brightest 'globular clusters', it's a great sight through a small telescope.

JUNE'S OBJECT

Mighty Jupiter boasts 79 moons – at the last count! But the most famous of **Jupiter's moons** are the four first observed by Galileo in 1610 with the newly invented telescope (though *really* sharp-sighted people can spot them with the unaided eye). The great scientist sketched the positions of these so-called 'Galilean moons' from night-to-night – convincing him that Copernicus was right in concluding that smaller objects (like planets) circle a larger body (the Sun).

It's fascinating to watch the antics of Galileo's moons – now you see them, now you don't! They are visible in binoculars, though best seen through a small-to-moderate telescope.

First off the mark is volcanic Io – it takes just 1.8 days to complete a circuit. Next is enigmatic Europa (thought to harbour an under-surface ocean) at 3.5 days. Ganymede – at 5268 kilometres across, the biggest moon in the Solar System (larger than the planet Mercury)

OBSERVING TIP

June is *the* month for the best Sun-viewing, with the chance of picking up dark sunspots blotting its fair face. But be careful. NEVER look at the Sun directly, with your naked eyes or – especially – with a telescope or binoculars: it could well blind you permanently. Project the Sun's image onto a piece of white card. Or you can use 'eclipse glasses' (meeting the ISO 12312-2 international safety standard) to reduce the Sun's radiation. Alternatively, buy solar binoculars or a solar telescope, with built-in filters that guarantee a safe and detailed view of the Sun's churning surface.

Nigel Bradbury took this shot on 8 July 2018 at 2.33 am. He used a Canon 6D camera, f/2.8 with a Sigma zoom lens (70–200 mm) at 104 mm. The exposure was 1.6 seconds.

– follows at 7.2 days, while heavily cratered Callisto brings up the rear at 16.7 days. You won't see any details on the moons, but it's fun to watch the action.

JUNE'S TOPIC: ANTIKYTHERA MECHANISM

Today, we have plenty of apps and computer software to calculate what's happening in the sky. But – amazingly – the world's first astronomical computer was built over 2000 years ago!

Discovered in a shipwreck off the Greek island of the same name, the Antikythera Mechanism is a corroded lump of bronze about the size of a shoebox. There's an obvious large gearwheel; and detailed X-ray analysis has shown 36 other gearwheels inside. Traces of writing on the outside suggest that the Antikythera Mechanism was constructed around 100–200 BC.

An astronomer would have worked this mechanical computer by turning a handle on the side, which rotated the internal gears. They drove pointers on a front dial, to show the position of the Sun and the Moon (and probably the planets) in the sky, and also the phases of the Moon. Another pointer on the back indicated when there would be eclipses of the Sun and the Moon.

JUNE'S PICTURE

Noctilucent clouds are the highest in the atmosphere: 80 kilometres up, they are right on the edge of space. They form above the Earth's poles, so you have to live between latitudes of 50 and 70 degrees to see them. These ghostly apparitions are best seen in the early summer months, but the big question is: what are they?

The clouds are clearly ice that's frozen onto tiny solid particles. These may be bits of cosmic dust, drifting down into the Earth's atmosphere. On the other hand, noctilucent clouds were first seen in 1885, soon after the eruption of the volcano Krakatoa – so they could be volcanic dust.

Nigel Bradbury explains how he captured this beautiful sight: 'With a view looking due north, Magpie Mine in the Peak District is ideal for spotting noctilucent clouds. I go up there every night between mid-May and late July, watching them descend in the north-west before midnight, and ascend in the north-east into the morning twilight.'

JUNE'S CALENDAR

SUNDAY	MONDAY	TUESDAY	WEDNESDAY	THURSDAY	FRIDAY	SATURDAY
	1 Moon near Spica	2 Moon near Spica	3	4 Moon near Antares; Mercury E elongation	5 8.12 pm Full Moon, near Antares	6
7	8 Moon near Jupiter and Saturn (am)	9 Moon near Jupiter and Saturn (am)	10	11	12	13 7.23 am Last Quarter Moon, near Mars and Neptune (am)
14 Mars near Neptune (am)	15	16	17	18	19 Moon occults Venus (am)	20 Summer Solstice
21 7.41 am New Moon, annular solar eclipse	22	23	24	25 Moon near Regulus	26	27
28 9.15 am First Quarter Moon	29 Moon near Spica	30				

SPECIAL EVENTS

- **8–9 June:** the bright 'star' near the Moon is Jupiter, with Saturn lying to its left (Chart 6a).
- **19 June:** the Moon moves right in front of Venus: this occultation takes place in daylight, so you'll need a telescope to see it. Just before sunrise, you'll spot the thin crescent Moon and the Morning Star (also a crescent at high magnification) together right on the horizon. Follow them as the Sun rises, and you'll see Venus disappear behind the Moon at 8.40 am (Chart 6b) and reappear at 9.40 am. YOU MUST TRACK THE MOON: DON'T SCAN THE TELESCOPE AROUND AT RANDOM OR YOU MAY ACCIDENTLY POINT IT AT THE SUN AND DAMAGE YOUR EYES.
- **20 June, 10.43 pm:** Summer Solstice. The Sun reaches its most northerly point in the sky, so today is Midsummer's Day, with the longest period of daylight and the shortest night.
- **21 June:** Seen from a narrow band from central Africa, across northern India to China, the Moon moves in front of the Sun, leaving a bright ring of the solar surface visible around its silhouette. People in most of Africa and Asia will witness a partial eclipse, but nothing of this annular eclipse is visible from the British Isles.

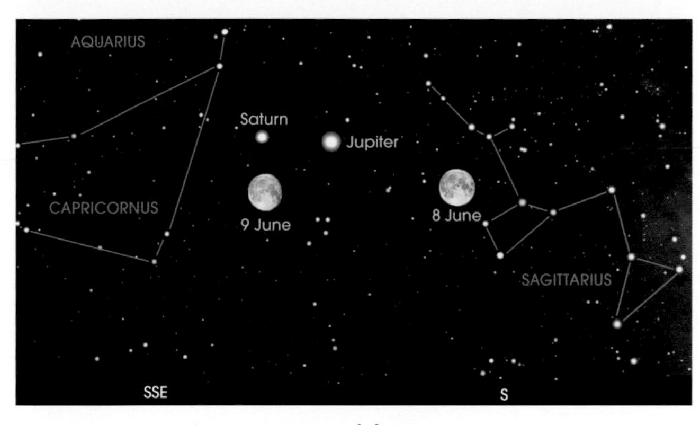

6a 8–9 June, 3 am. Saturn, Jupiter and the Moon.

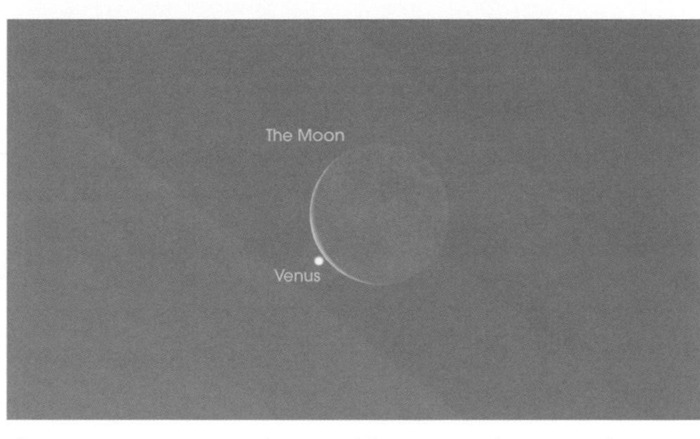

6b 19 June, 8.35 am. Binocular view of the Moon occulting Venus during daylight.

- **Mercury** skulks down on the north-western horizon, at greatest elongation from the Sun on 4 June and setting at 11 pm. From magnitude +0.3 at the start of the month, it fades quickly and sinks into the twilight glow by the middle of June.
- Giant planet **Jupiter** is rising in the south-east around 11 pm, at a brilliant magnitude −2.6 in Sagittarius. **Saturn** lies just five degrees to its left in Capricornus, 15 times fainter at magnitude +0.3 and rising 20 minutes later.
- **Mars** rises around 1.30 am. The Red Planet brightens noticeably during the month as the Earth speeds towards it, increasing from magnitude 0.0 to −0.5 as it moves from Aquarius into Pisces.
- Aquarius also hosts **Neptune**, which is 1.5 degrees above Mars on the mornings of 13 and 14 June. The most distant planet glows at magnitude +7.9 and clears the horizon about 1 am.
- **Uranus** (magnitude +5.9) lies in Pisces and rises around 2.30 am.
- **Venus** roars into dawn skies mid-month, as a stunning Morning Star rising in the north-east and blazing at magnitude −4.3. The narrowest crescent Moon lies to the right of Venus on the morning of 19 June – use binoculars to look low on the horizon around 4.30 am – and the Moon occults Venus later in the morning (see Special Events and Chart 6b). By the end of June, Venus is rising as early as 3 am.

JUNE'S PLANET WATCH

WEST

- The sky at 11 pm in mid-July, with Moon positions at three-day intervals either side of Full Moon.
- The star positions are also correct for midnight at the beginning of July, and 10 pm at the end of the month.
- The planets move slightly relative to the stars during the month.

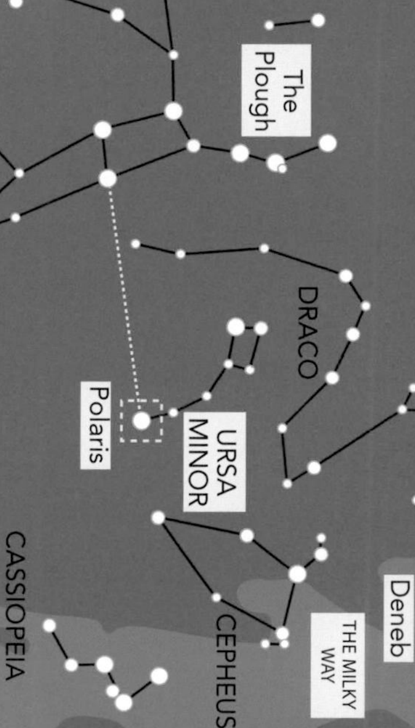

NW
VIRGO
LEO
The Sickle
CANES VENATICI
BOÖTES
URSA MAJOR
HERCULES
LOOKING NORTH
AURIGA
Capella
Zenith
Deneb
THE MILKY WAY
CYGNUS
PERSEUS
Algol
PEGASUS
Square of Pegasus
TRIANGULUM
ANDROMEDA
NE
PISCES

EAST

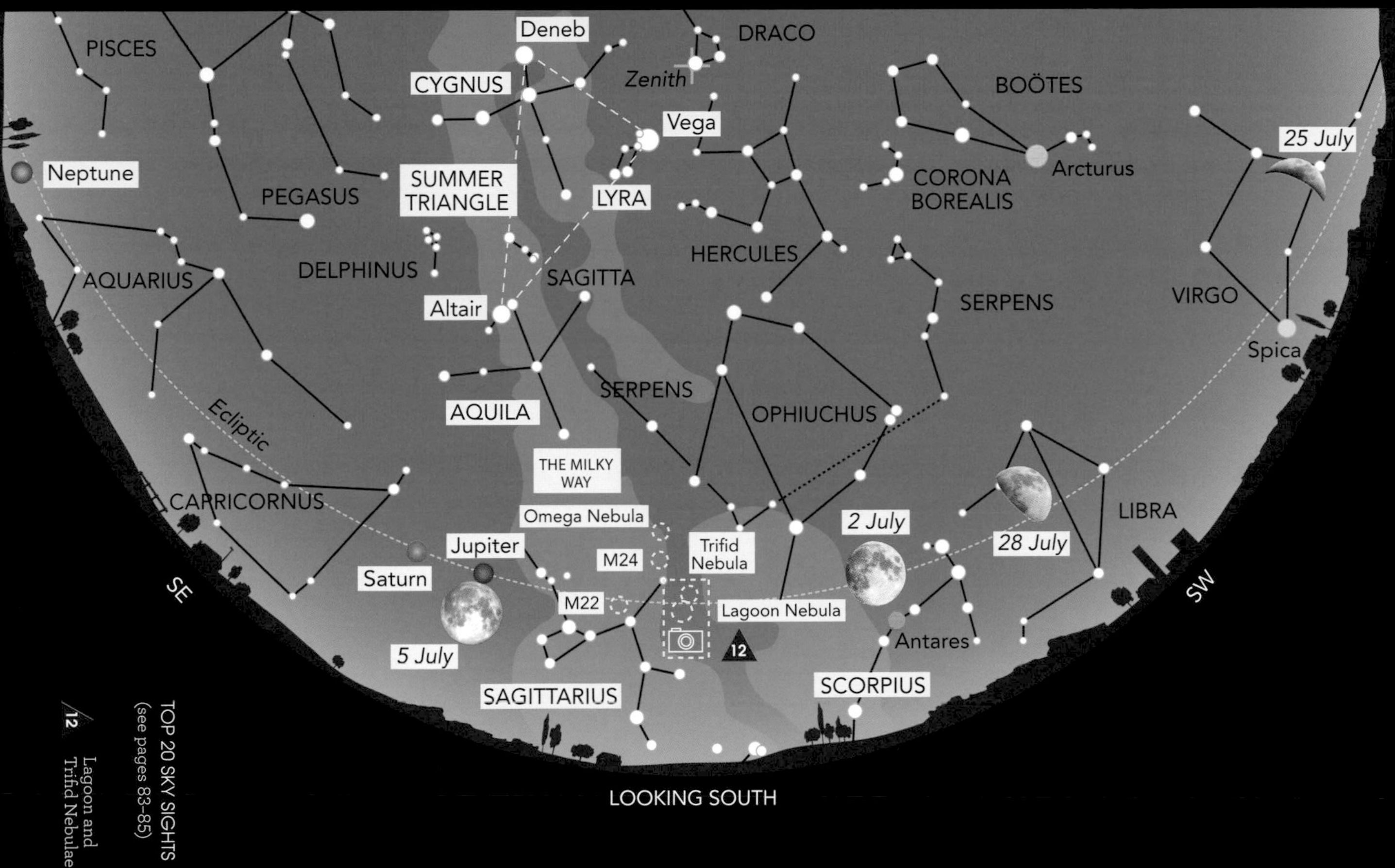

TOP 20 SKY SIGHTS
(see pages 83–85)

12 Lagoon and Trifid Nebulae

These short summer nights are illuminated by a pair of brilliant planets – **Jupiter** and **Saturn** – hobnobbing down in the south among the stars of **Sagittarius** and **Scorpius**, which are embedded in the glorious heart of the **Milky Way**. Higher in the sky, the prominent **Summer Triangle** is composed of **Vega**, **Deneb** and **Altair**, the brightest stars in **Lyra**, **Cygnus**, and **Aquila**.

JULY'S CONSTELLATION

Just to the right of Jupiter – the brightest object in the night sky – you'll find a constellation that's shaped rather like a teapot, with the handle to the left; and the spout to the right.

To the ancient Greeks, the star-pattern of **Sagittarius** represented an archer, with the torso of a man and the body of a horse. The 'handle' of the teapot represents his upper body; the curve of three stars to the right his bent bow; while the end of the spout is the point of the arrow, aimed at Scorpius, the fearsome celestial scorpion.

Sagittarius is rich in nebulae and star clusters. If you have a clear night (and preferably from a southern latitude), sweep Sagittarius with binoculars for some fantastic sights. Above the spout lies the wonderful **Lagoon Nebula** – a region of starbirth that's visible to the naked eye on clear nights. Nearby is the small-but-perfectly-formed three-lobed **Trifid Nebula** (telescope required). Between Sagittarius and **Aquila**, you'll find a bright patch of stars in the **Milky Way**, catalogued as **M24**. Raise your binoculars higher to spot another star-forming region, the **Omega Nebula**.

Finally, on a very dark night you may spot a fuzzy patch, above and to the left of the Teapot's lid. This is the globular cluster **M22**, a swarm of almost a million stars that lies 10,600 light years away.

OBSERVING TIP

This is the month when you really need a good, unobstructed horizon to the south, for the best views of the glorious summer constellations of Scorpius and Sagittarius. They never rise high in temperate latitudes, so make the best of a southerly view – especially over the sea – if you're away on holiday. A good southern horizon is also best for views of the planets, because they rise highest when they're in the south.

JULY'S OBJECT

The Pole Star – **Polaris** – is a surprisingly shy animal, coming in at the modest magnitude of +2.0. Find it by following the two end stars of the **Plough** (see Star Chart) in **Ursa Major**, the Great Bear. Polaris lies at the end of the tail of the Little Bear (**Ursa Minor**), and it pulsates in size, making its brightness vary slightly over a period of four days. But its importance centres on the fact that Earth's north pole points towards Polaris, so we spin 'underneath' it. It remains almost stationary in the sky, and acts as a fixed point for both astronomy and navigation. But the Earth's axis swings slowly around – a phenomenon called precession (see September's Topic) – so our 'pole stars' change with time.

JULY'S TOPIC: RINGWORLDS

Rings around planets – and even moons – are common in the outer Solar System.

The most famous rings of all belong to **Saturn**. When we show the planet to people through a telescope, they can't believe that it's real – 'You must have hung a model in front of the lens!'

Real they are; and so wide that they would almost stretch to the Moon if they girdled Earth. Saturn's rings are made of billions of ice fragments, ranging in size from ice cubes to refrigerators. They were almost certainly formed by the break-up of a moon of Saturn that got torn apart a hundred million years ago – that's recent in cosmic history.

Jupiter, Uranus and **Neptune** all have rings: but you need a spaceprobe to see them. Jupiter's are fairly wimpish. They're dark, and made of dust particles shed by its moons. Uranus has 13 rings – also very dark. These are made of ice, but coated with particles of sooty dust. Neptune is home to a family of five rings – all dark and dusty.

There are tantalising hints that some moons of the outer planets are circled by rings. And the dwarf planet Haumea, at the edge of the Solar System, is also a ringworld.

JULY'S PICTURE

A beautiful pair of nebulae located in Sagittarius, taken by Pete Williamson. Below is the Lagoon Nebula, above is the Trifid. The interloper at top left is the planet Saturn. The dark dust-lanes in the nebulae are made of dust shed by old stars, which creates rocky planets like the Earth.

Pete Williamson, based in Shropshire, captured this stunning image in August 2018 by remote-observing with a Takahashi 106 mm telescope at Siding Spring, Australia. Pete used three filters at different exposures: red, green and blue (all at 5 × 300 seconds). He processed the images in PixInsight 1.8, and combined them in Photoshop.

JULY'S CALENDAR

SUNDAY	MONDAY	TUESDAY	WEDNESDAY	THURSDAY	FRIDAY	SATURDAY
			1	2 Moon near Antares	3	4 Earth at aphelion
5 5.44 am Full Moon near Jupiter	6 Moon near Saturn and Jupiter	7	8	9	10	11 Venus near Aldebaran (am)
12 Moon near Mars (am); Venus near Aldebaran (am)	13 0.29 am Last Quarter Moon	14 Jupiter opposition	15	16 Moon below the Pleiades (am)	17 Moon near Venus, Aldebaran, Hyades and Pleiades (am)	18 Moon occults Crab Nebula (am)
19	20 6.33 pm New Moon; Saturn opposition	21	22 Moon near Regulus; Mercury W elongation	23	24	25
26 Moon near Spica	27 1.32 pm First Quarter Moon	28	29 Moon near Antares	30	31	

SPECIAL EVENTS

- **4 July, 12.34 pm:** the Earth is furthest from the Sun (aphelion), at 152 million km.
- **5 July:** look out for a stunning sight as the Full Moon passes below Jupiter near maximum brilliance (Chart 7a).
- **6 July:** the Moon lies below Saturn, with Jupiter to the right (Chart 7a).
- **11–12 July:** Venus passes above Aldebaran (see Planet Watch).
- **12 July:** Mars lies above the Moon in the early morning.
- **14 July:** Jupiter is opposite to the Sun in the sky. Since the planets' orbits are not circular, we are closest to Jupiter on 15 July, at 619 million km.
- **16 July:** the Moon lies below the Pleiades in the small hours, with dazzling Venus to the left (Chart 7b).
- **17 July:** an unforgettable dawn spectacle as the Moon and Venus converge near Aldebaran (see Planet Watch).
- **18 July:** with a clear north-eastern horizon and a telescope, watch the crescent Moon move in front of the Crab Nebula, starting at 3.25 am and ending at 4.25 am (Chart 7b).
- **20 July:** Saturn is opposite to the Sun in the sky; it's closest to Earth the next day, 1346 million km away.
- **This month,** a flotilla of spacecraft heads towards Mars: Europe's ExoMars Rover, the NASA Mars 2020 rover, a Chinese lander and rover, the Hope mission from the United Arab Emirates – and perhaps private missions, too.

JULY'S PLANET WATCH

7a 4–6 July, 11.30 pm. Saturn, Jupiter and the Full Moon.

7b 16–18 July, 3 am. Moon, Venus, Aldebaran, Hyades and Crab Nebula occultation.

- **Jupiter** is king of the night, at opposition on 14 July, and blazing low in the southern sky all night long in Sagittarius. At magnitude –2.7, the giant planet far outshines any star.
- Just to the left of Jupiter you'll find **Saturn**, ten times fainter at magnitude +0.1 and also visible all night in Sagittarius: the ringworld reaches opposition on 20 July.
- **Mars** is rising in the east around midnight, intermediate in brightness between Jupiter and Saturn. During July, the Red Planet brightens from magnitude –0.5 to –1.1 as it travels through Pisces.
- Dim **Neptune** (magnitude +7.8) rises in Aquarius about 11 pm. Its near-twin **Uranus** is rising around 0.30 am in Aries, and shines at magnitude +5.8.
- Brilliant **Venus** – magnitude –4.4 – appears before the Sun in the morning sky, rising about 2.30 am. During the first half of the month, the Morning Star lies among the stars of the Hyades (best viewed in binoculars) and on the mornings of 11 and 12 July it passes just above Aldebaran.
- Just before dawn on 17 July, you'll find Venus just below the thin crescent Moon. The star to the right is Aldebaran: look closely (best in binoculars) and you'll see the stars of the Hyades swarming near the Moon. The whole show is set off by the Pleiades to the top right (Chart 7b).
- During the last week of July, look for **Mercury** on the horizon well to the lower left of Venus and reaching western elongation on 22 July. At magnitude –0.5, it's rising just before 4 am.

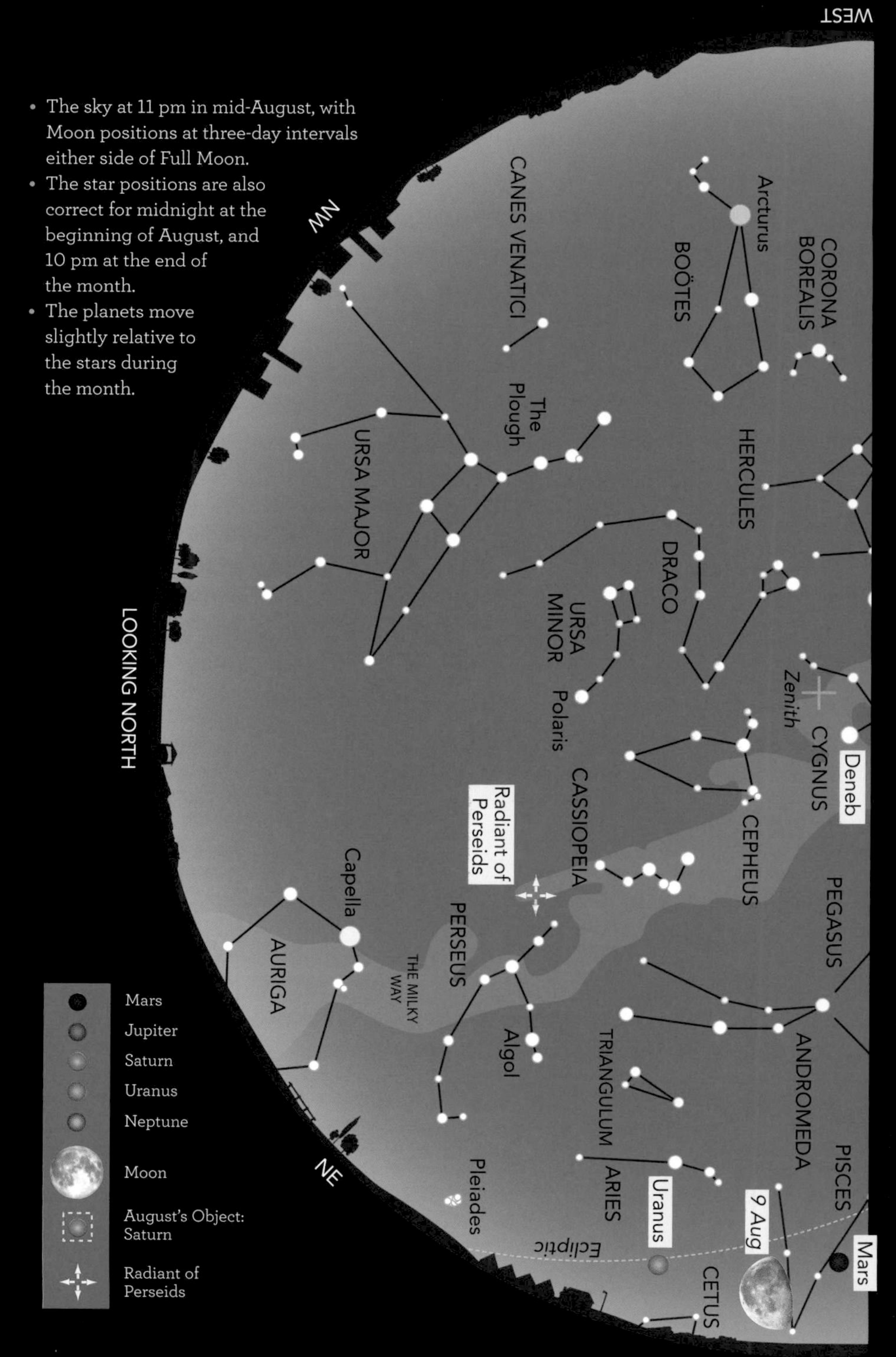

- The sky at 11 pm in mid-August, with Moon positions at three-day intervals either side of Full Moon.
- The star positions are also correct for midnight at the beginning of August, and 10 pm at the end of the month.
- The planets move slightly relative to the stars during the month.

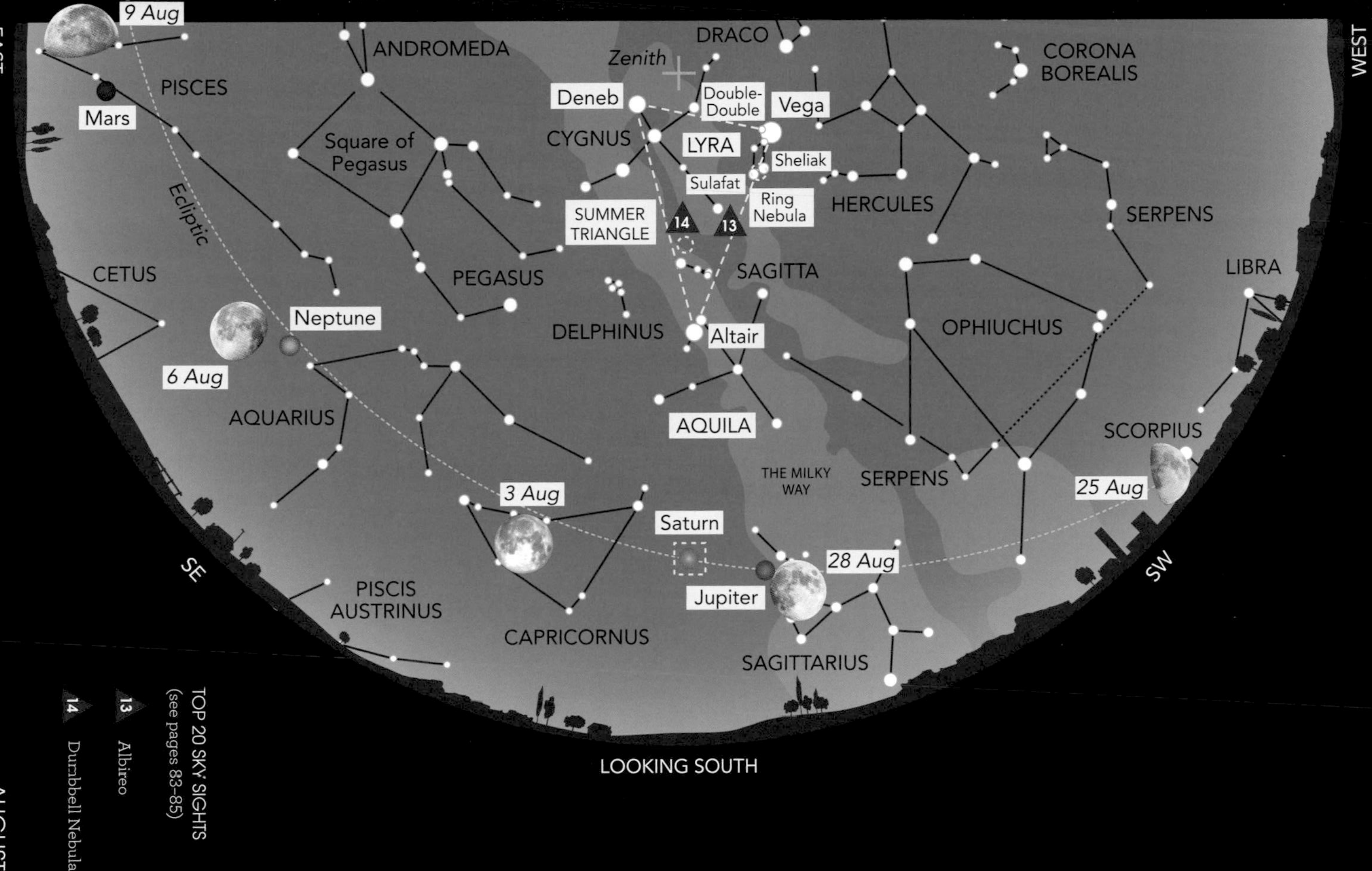

TOP 20 SKY SIGHTS
(see pages 83–85)

13 Albireo

14 Dumbbell Nebula

AUGUST

If you've wanted to spot all eight planets of the Solar System in a single night, take advantage of this month's short and warm hours of darkness. You can't miss **Jupiter**, **Saturn** and **Mars** in the evening sky, and good binoculars will winkle out fainter **Uranus** and **Neptune**. Party on till 4.30 am, and you can add Venus and Mercury to your bag. Oh yes, and to add the final planet – just look beneath your feet.

AUGUST'S CONSTELLATION

Lyra is small but perfectly formed. Shaped like a Greek lyre, it's dominated by brilliant white **Vega**, the fifth-brightest star in the sky. Just 25 light years away – a near neighbour in the Cosmos – Vega is surrounded by a disc of dust that has probably given birth to baby planets.

Next to Vega is the **Double-Double**, a quadruple star known officially as epsilon Lyrae. Keen-sighted people can separate the pair, but you'll need a small telescope to find that each star is itself double.

The gem of Lyra lies between the two end stars of the constellation, **Sheliak** and **Sulafat**. The **Ring Nebula** needs a serious telescope (it's nearly ninth magnitude and little larger than Jupiter in apparent size), and is a wonderful example of a planetary nebula. Named by William Herschel, famed for his discovery of the planet Uranus, planetary nebulae look at first glance like dim, distant worlds. But in fact the Ring Nebula is a ghostly star corpse that has puffed away its atmosphere: the end of the road for a star like the Sun.

OBSERVING TIP

Don't think that you need a telescope to bring the heavens closer. Binoculars are excellent – and you can fling them into the back of the car at the last minute. For astronomy, buy binoculars with large lenses coupled with a modest magnification. An ideal size is 7 × 50, meaning that the magnification is seven times, and that the diameter of the lenses is 50 millimetres. They have good light grasp, and the low magnification means that they don't exaggerate the wobbles of your arms too much. Even so, it's best to rest your binoculars on a wall or a fence to steady the image.

AUGUST'S OBJECT

Saturn, just past its closest this year, is livening up the southern constellation of Sagittarius (the Archer). It's a glorious sight through a small telescope: the world looks surreal, like an exquisite model hanging in space.

The planet is famed for its huge engirdling appendages: its rings would stretch nearly all the way from the Earth to the Moon. The rings are made of myriad ice particles, and the Cassini spaceprobe found that some of them are clumping together into moonlets. And that's just the beginning of Saturn's larger family. It has at least 62 moons, including Titan (visible through a small telescope) which boasts lakes of liquid methane and ethane, and possibly active volcanoes. The icy moon Enceladus is spewing salty water into space, while Dione has traces of oxygen in its thin atmosphere. These

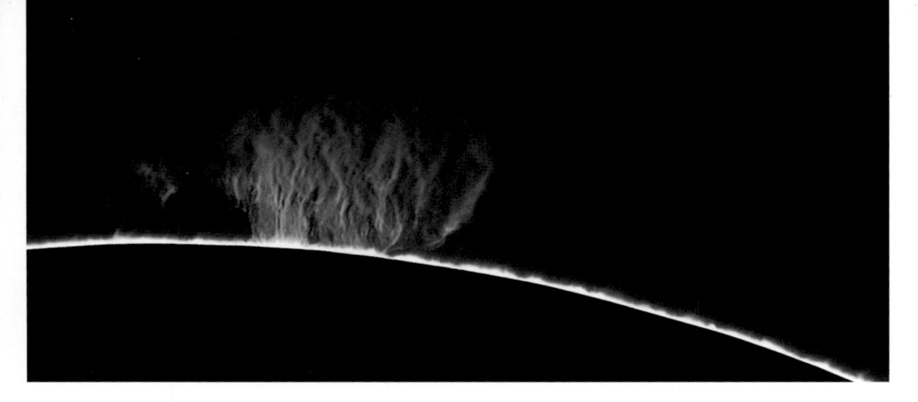

Alastair Woodward used a Sky-Watcher 120 mm achromat refractor (f/8.33) with a Daystar Quark Chromosphere hydrogen-alpha eyepiece. His camera was a FLIR Blackfly 2.3 MP Mono GigE CMOS. The exposure was 30 seconds at 42 frames per second. Alastair stacked 70 per cent of the best frames, and used image processing for sharpening.

discoveries raise the intriguing possibility of primitive life on Saturn's moons . . .

Saturn itself is second only to Jupiter in size. But its density is so low that – were you to plop it in an ocean – it would float. Saturn has a ferocious spin rate, turning once in 10 hours 33 minutes, and its winds roar at speeds of up to 1800 kilometres per hour.

AUGUST'S TOPIC: MAGNETIC FIELDS

Gravity is the main force shaping the Solar System, galaxies and the wider Universe. But there's another influence at play in the Cosmos, with the power to protect and destroy: the invisible tendrils of magnetism.

The Sun's magnetism holds clouds of hot gas in thrall (see Picture): when its magnetic loops short-circuit, they fling lethal streams of electrically charged particles across the Solar System. Distant dying stars create magnetic fireballs – supernova remnants – that shoot out deadly particles called cosmic rays.

Fortunately, our planet's core generates a powerful magnetic field that wraps the Earth in an invisible cocoon (the magnetosphere), shielding us from solar eruptions and cosmic rays. We can see some of this energy being dissipated harmlessly in the glorious light-show of an aurora (see February's Picture).

Further out in space, magnetic fields shape the clouds where stars are born – including the famous dark columns of the Pillars of Creation – and help to marshal the swirling gas around young stars, to form an orderly system of planets.

And spinning black holes within distant galaxies can wind up colossal magnetic forces, which channel intense 'jets' of radiation in narrow beams of lethal energy that can reach a million light years in length.

AUGUST'S PICTURE

This fiery blaze of light is a solar prominence, soaring thousands of kilometres above the edge of the Sun. Prominences appear above groups of sunspots, and are a sensational sign of our local star's magnetic activity. You can see prominences at a total eclipse of the Sun, or through specialised photographic equipment (as captured by Alastair Woodward here from his back garden in Derby). These dramatic outbursts can last for weeks.

AUGUST'S CALENDAR

SUNDAY	MONDAY	TUESDAY	WEDNESDAY	THURSDAY	FRIDAY	SATURDAY
30	31					1 Moon near Jupiter
2 Moon near Saturn	3 4.58 pm Full Moon	4	5	6	7	8 Moon near Mars
9 Mercury amid Praesepe (am)	10	11 5.44 pm Last Quarter Moon	12 Perseids	13 Perseids (am); Venus W elongation; Moon near Aldebaran (am)	14	15 Moon near Venus (am)
16 Moon near Venus (am)	17	18	19 3.41 am New Moon	20	21	22 Moon near Spica
23	24	25 6.57 pm First Quarter Moon near Antares	26 Moon near Antares	27	28 Moon near Jupiter	29 Moon near Saturn

SPECIAL EVENTS

- **1 August:** the bright 'star' located above the Moon is Jupiter, while the fainter planet Saturn can be seen to the left (Chart 8a).
- **2 August:** Saturn lies near the Moon, with the brilliant giant planet Jupiter to the right (Chart 8a).
- **8 August:** the Moon is close to Mars tonight.
- **9 August:** low in the dawn twilight to the north-east, use binoculars or a telescope to spot Mercury in front of the Praesepe star cluster (below Castor and Pollux).
- **Night of 12/13 August:** maximum of the Perseid meteor shower. The Earth runs into debris from Comet Swift-Tuttle, which burns up in the atmosphere. The shooting stars are most prolific in the early hours, but this year the show is spoilt by the Moon rising at midnight.
- **13 August:** a beautiful tableau in the early hours, with the crescent Moon in the Hyades above bright Aldebaran, with the Pleiades to the upper right and dazzling Venus to the lower left (Chart 8b).
- **15–16 August:** the crescent Moon forms a striking pair with Venus just before dawn (Chart 8b).
- **28 August:** Jupiter lies close to the Moon, with Saturn to the left (Chart 8c).
- **29 August:** the Moon passes under Saturn, while brighter Jupiter lies to the right (Chart 8c).

AUGUST'S PLANET WATCH

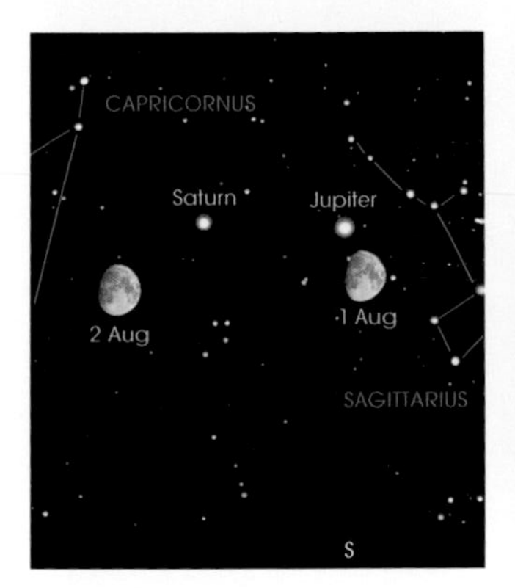

8a *1–2 August, 11 pm. Moon with Saturn and Jupiter.*

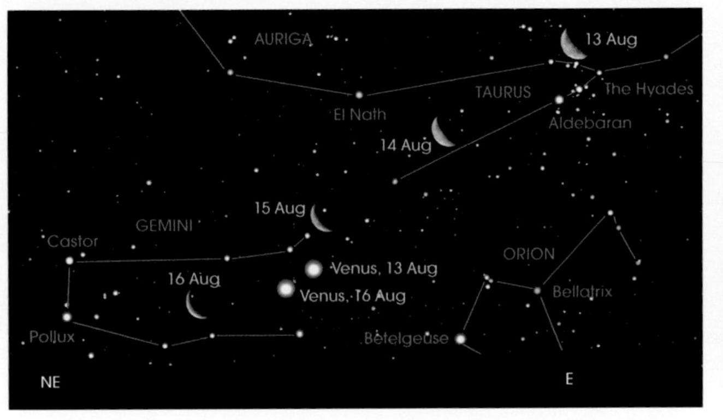

8b *13–16 August, 3 am. Venus and the Moon with Castor, Pollux, Aldebaran and the Hyades.*

8c *28–29 August, 11 pm. Moon with Saturn and Jupiter.*

• Though they are both fading as the Earth draws away, the giant planets **Jupiter** and **Saturn** still dominate the skies to the south, both in Sagittarius and setting about 3 am. Jupiter, to the right, is much the brighter at magnitude –2.6. Fifteen times fainter at magnitude +0.2, Saturn still outshines the surrounding stars. Use binoculars to spot Jupiter's four large moons, and a telescope to view Saturn's biggest moons and its iconic rings.

• **Mars** is rising in the east around 10.30 pm in Pisces. The Red Planet brightens from magnitude –1.1 to –1.8 as the Earth speeds towards it.

• You'll need binoculars or a telescope to spot Neptune, at magnitude +7.8 in Aquarius and rising about 9 pm. **Uranus** is just on the borderline of naked-eye visibility at magnitude +5.7. It rises around 10.30 pm in Aries.

• **Venus** is growing ever more conspicuous in the morning, as it draws upwards into darker skies. Blazing at magnitude –4.3, the Morning Star is rising about 2 am. Through good binoculars or a small telescope, you'll see Venus change from a crescent to a half-lit globe as the month progresses.

• During the first half of August, look low in north-eastern morning twilight to spot elusive **Mercury**. At magnitude –1.2, the innermost planet rises around 4.20 am and lies below fainter Castor and Pollux. On 9 August, it passes in front of Praesepe (see Special Events).

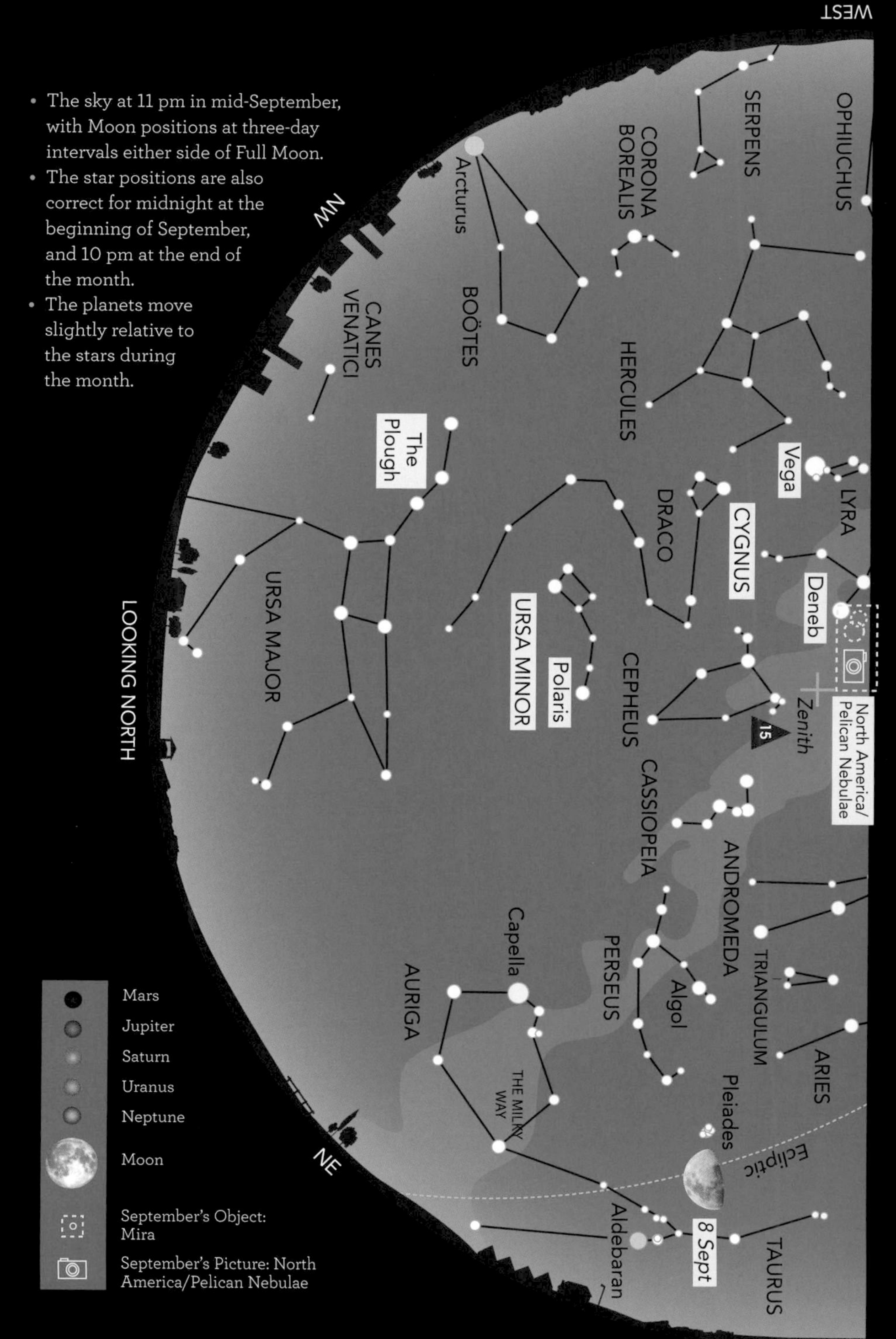

- The sky at 11 pm in mid-September, with Moon positions at three-day intervals either side of Full Moon.
- The star positions are also correct for midnight at the beginning of September, and 10 pm at the end of the month.
- The planets move slightly relative to the stars during the month.

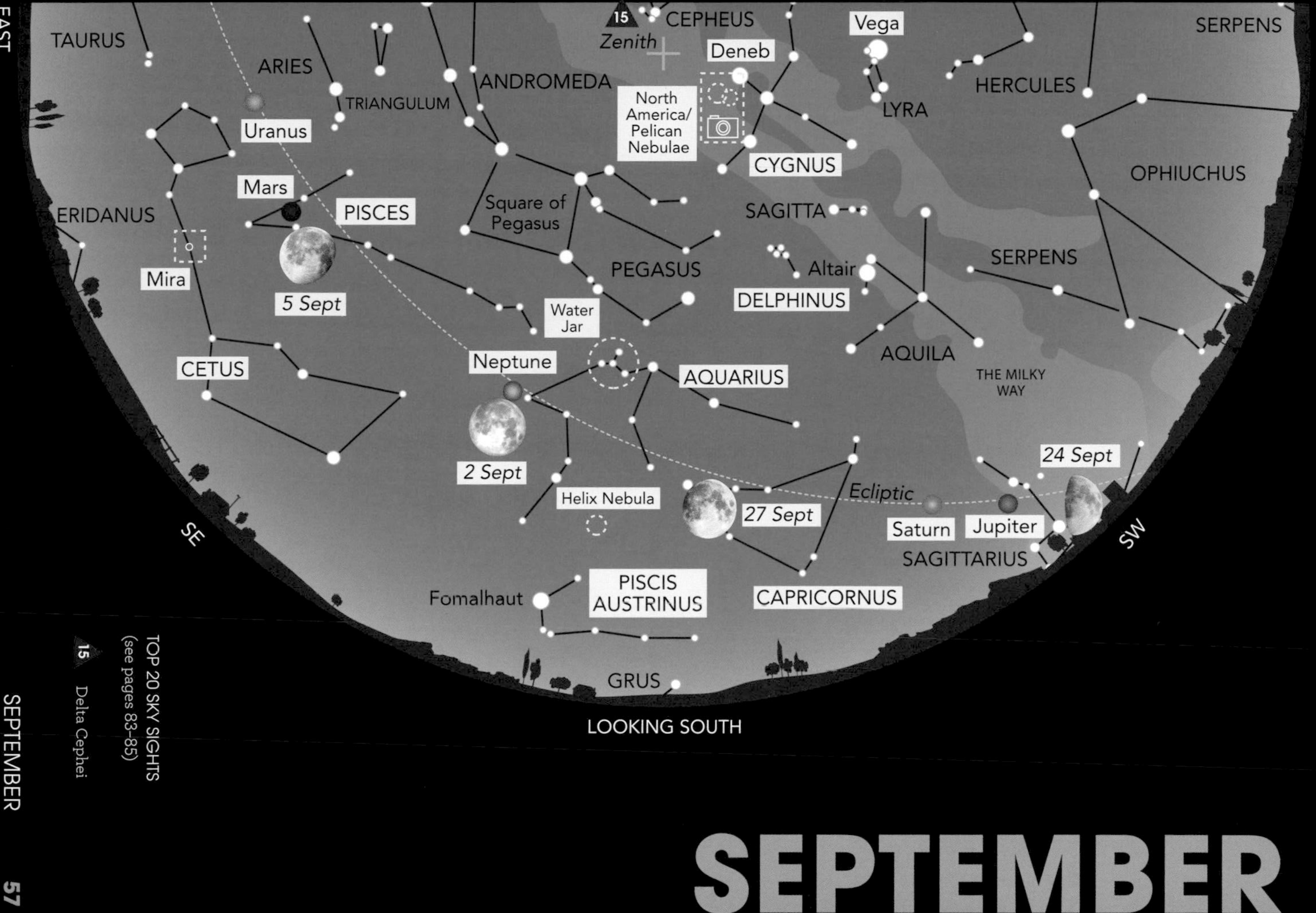

TOP 20 SKY SIGHTS
(see pages 83–85)

15 Delta Cephei

SEPTEMBER

There's a host of dim watery constellations in this month's celestial tableau: in the swim with our featured star pattern of **Aquarius**, you can find **Cetus** (the Sea Monster), **Capricornus** (the Sea Goat), **Pisces** (the Fishes), **Piscis Austrinus** (the Southern Fish) and **Delphinus** (the Dolphin). The night is book-ended by the planets **Jupiter** – brilliant after sunset – and Venus heralding the dawn.

SEPTEMBER'S CONSTELLATION

Although hardly one of the most spectacular constellations, **Aquarius** (the Water Carrier) has a pedigree stretching back to antiquity. The ancient Babylonians possibly associated this zone of the heavens with water because the Sun passed through it during the rainy season. The Greeks depicted Aquarius as a man pouring from a **Water Jar** (the central group of four faint stars), the liquid splashing downwards onto **Piscis Austrinus**.

Aquarius boasts one of the most glorious sky-sights in long-exposure images; it's visible as a faint celestial ghost in binoculars or through a small telescope. Half the diameter of the Full Moon, the **Helix Nebula** is a star in its death throes. It's a 'planetary nebula' – and, at 700 light years away, one of the nearest. An aged red giant star has puffed off its unstable, distended atmosphere to form a beautiful shroud around its collapsed core – a white dwarf star that will gradually fade away to become a cold black cinder.

SEPTEMBER'S OBJECT

In another fishy constellation close to Aquarius – **Cetus** – we find an extraordinary star. It was first observed in 1596 by the German astronomer David Fabricius. He thought it was an exploding star, because it flared up and then vanished. But – less than a year later – it was back. Fabricius had discovered the first variable star, and called it **Mira** – Latin for 'the wonderful'.

Mira is a foretaste of what will befall our Sun, seven billion years hence. It has swollen into a bloated red giant star: so vast that – were it placed in our Solar System – it would engulf all the planets out to the asteroid belt.

With little gravitational control over its girth, Mira swells and shrinks over a period of 332 days, rising to a bright magnitude +2 at maximum, and only magnitude +10 at minimum, when you need a medium-sized telescope to see it at all.

Mira is charging through space at a reckless rate, and – as a result – the unstable star is trailing a tail of gas 13 light years long. But this isn't all bad news. The elements in Mira's tail have the power to kick-start new stars and planets into being born.

SEPTEMBER'S TOPIC: PRECESSION

People often ask us, 'Is the Pole Star the brightest star in the sky?' – and they're surprised to hear it's not. The Pole Star is whatever star happens to lie overhead if you stand at the North Pole: a star here stays in the same place in the heavens as the Earth spins round; and from anywhere in the northern hemisphere, it always lies due north.

Simon Hudson's equipment was an Altair Astro 70 EDQ-R telescope, coupled to a QHY9s CCD camera. To capture the spooky colours, he used Baader narrowband filters. He brought out the individual elements with different exposures: 24 × 600-second exposures for hydrogen-alpha; 17 × 600-second exposures for oxygen III. The final image was processed using Pixinsight and Photoshop.

It's a matter of chance whether there's a bright star in that direction at all. The southern hemisphere doesn't have a significant pole star: Sigma Octantis, barely visible to the unaided eye, hovers some distance from prime pole position.

In the northern half of the Earth, we're fortunate that there's a moderately bright star close to the sky's north pole. At a magnitude of +2.0, **Polaris** – in the tail of the Little Bear (**Ursa Minor**) – is about the same brightness as the stars in the familiar **Plough**.

But the Earth's axis is gradually swinging round, over a period of 26,000 years, like a toy spinning-top about to topple over – a process called precession. As a result, Polaris will be closest to true north on 24 March 2100. Hang on for 12,000 years or so, and we will be blessed with a dazzling Pole Star, in the shape of the brilliant star **Vega**.

SEPTEMBER'S PICTURE

Simon Hudson captured this haunting image of a nebula complex near **Deneb** in **Cygnus**. On the left is the aptly named **North America Nebula**; below, to the right, is the **Pelican Nebula**. Both are part of the same giant gas cloud, lying 1600 light years away, but split by a massive band of dust that creates a cosmic 'gulf of Mexico'. This will be a formation site for future stars and planets.

OBSERVING TIP

It's best to view your favourite objects when they're well clear of the horizon. If you observe them low down, you're looking through a large thickness of the atmosphere – which is always shifting and turbulent. It's like trying to observe the outside world from the bottom of a swimming pool! This turbulence makes the stars appear to twinkle. Low-down planets also twinkle – although to a lesser extent, because they subtend tiny discs, and aren't so affected.

SEPTEMBER'S CALENDAR

SUNDAY	MONDAY	TUESDAY	WEDNESDAY	THURSDAY	FRIDAY	SATURDAY
		1	2 6.22 am Full Moon	3	4	5 Moon near Mars
6 Moon very near Mars (am)	7	8 Moon near the Pleiades	9	10 10.25 am Last Quarter Moon, near Aldebaran	11 Neptune opposition	12
13 Venus near Praesepe (am); Moon near Castor and Pollux (am)	14 Moon and Venus near Praesepe (am)	15	16	17 12.00 noon New Moon	18	19
20	21	22 Autumn Equinox; Moon near Antares	23 Moon near Praesepe (am)	24 2.55 am First Quarter Moon, near Jupiter	25 Moon near Saturn	26
27	28	29	30			

SPECIAL EVENTS

- **Night of 5/6 September:** the waning Moon is very close to brilliant red Mars (Chart 9a).
- **11 September:** Neptune is opposite to the Sun in the sky and at its closest to Earth this year, 4327 million km away.
- **13 September:** Venus lies near Praesepe, with the Moon to the right.
- **14 September:** a glorious sight in the morning sky, as the crescent Moon passes above dazzling Venus. Look more closely (preferably with binoculars), and you'll spot Praesepe (the Beehive star cluster) lying between them (Chart 9b).
- **22 September, 2.30 pm:** nights become shorter than days as the Sun moves south of the Equator at the Autumn Equinox.
- **24 September:** the bright 'star' near the First Quarter Moon is giant planet Jupiter.
- **25 September:** the Moon passes under Saturn, with Jupiter to the right.

Praesepe

SEPTEMBER'S PLANET WATCH

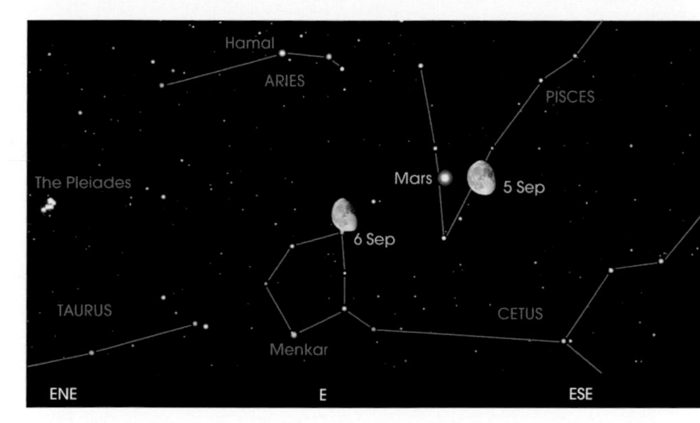

9a *5–6 September, 11 pm. Mars and the Moon.*

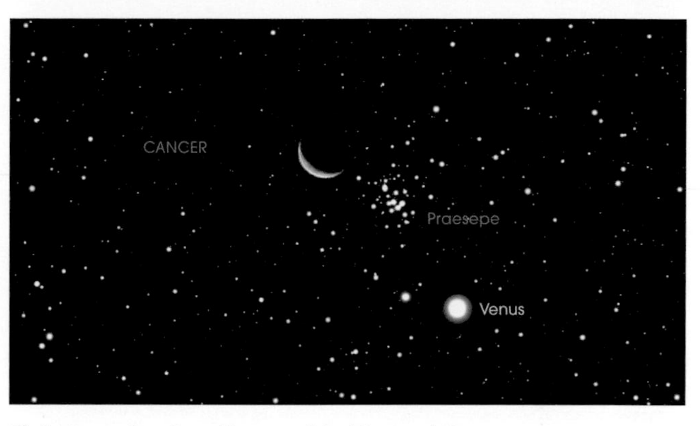

9b *14 September, 5 am. Venus and the Moon with Praesepe.*

- In the south-west, mighty **Jupiter** is lording it over the faint stars of Sagittarius. Shining at magnitude −2.5, the giant planet sets soon after midnight.
- To its left, **Saturn** – also in Sagittarius – is at a fainter magnitude +0.4, but still outshining the surrounding stars. The ringworld is setting about 1 am.
- Across in the east, **Mars** is putting on a strong challenge to Jupiter. As the Earth draws nearer to the Red Planet during September, its brightness increases from magnitude −1.8 to −2.5. You'll find Mars in Pisces, rising around 8.30 pm.
- **Neptune** is at its closest to the Earth on 11 September (see Special Events), although even then it's at a paltry magnitude +7.8 and only visible in binoculars or a telescope. The most distant planet lies in Aquarius, and is visible all night long.
- Its slightly brighter twin, **Uranus**, shines at magnitude +5.7 in Aries, and rises about 8.30 pm.
- **Venus** is a magnificent sight in the wee small hours. Rising as early as 2.30 am, the Morning Star blazes at magnitude −4.1, and it's the more resplendent because we see it in a totally dark sky. Venus starts the month below Castor and Pollux, and passes near the star cluster Praesepe on the mornings of 13 and 14 September – on the latter date, it forms a striking tableau with the crescent Moon (see Special Events and Chart 9b).
- **Mercury** is too close to the Sun to be visible this month.

- The sky at 11 pm in mid-October, with Moon positions at three-day intervals either side of Full Moon.
- The star positions are also correct for midnight at the beginning of October, and 9 pm at the end of the month (after the end of BST).
- The planets move slightly relative to the stars during the month.

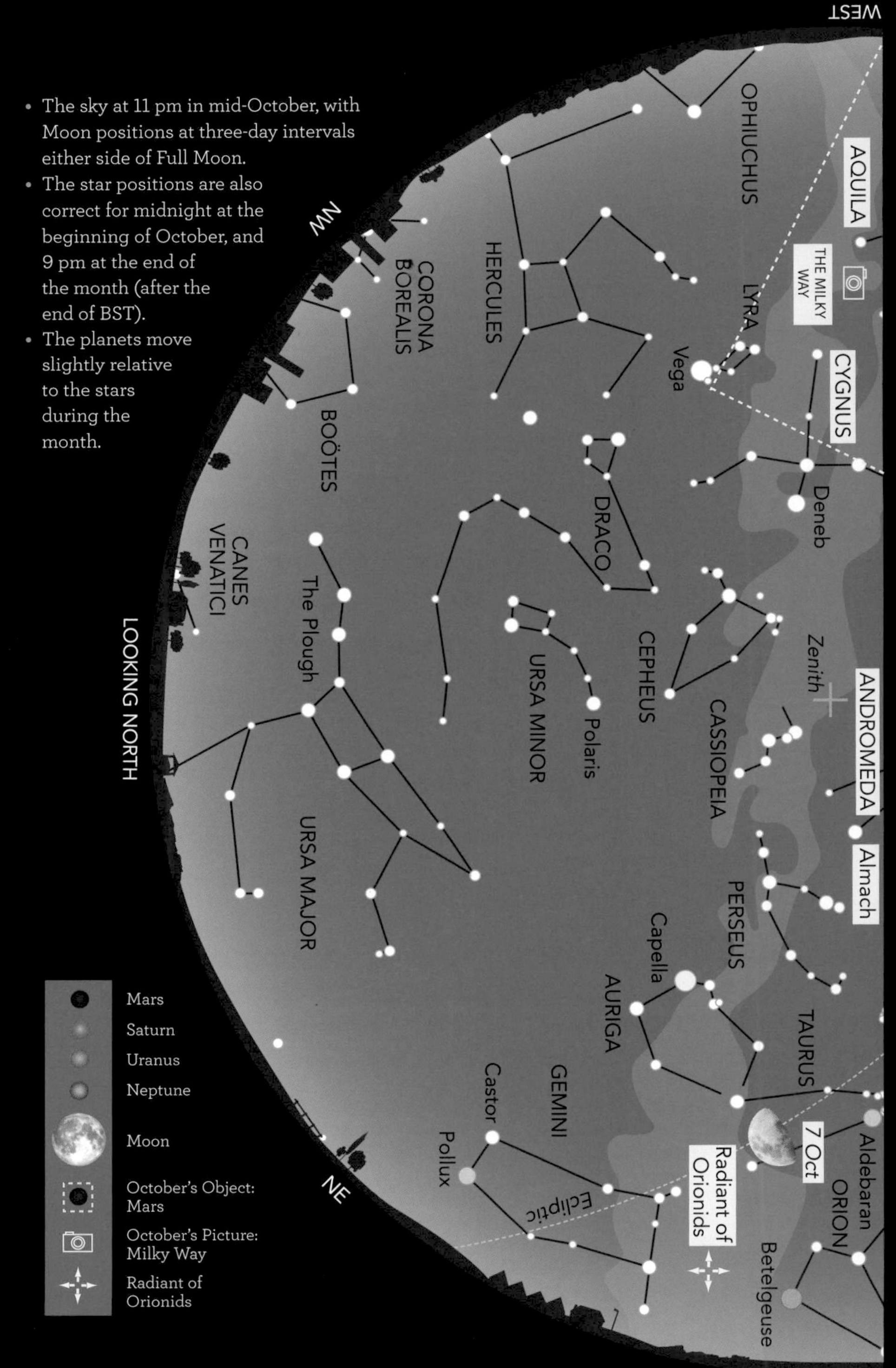

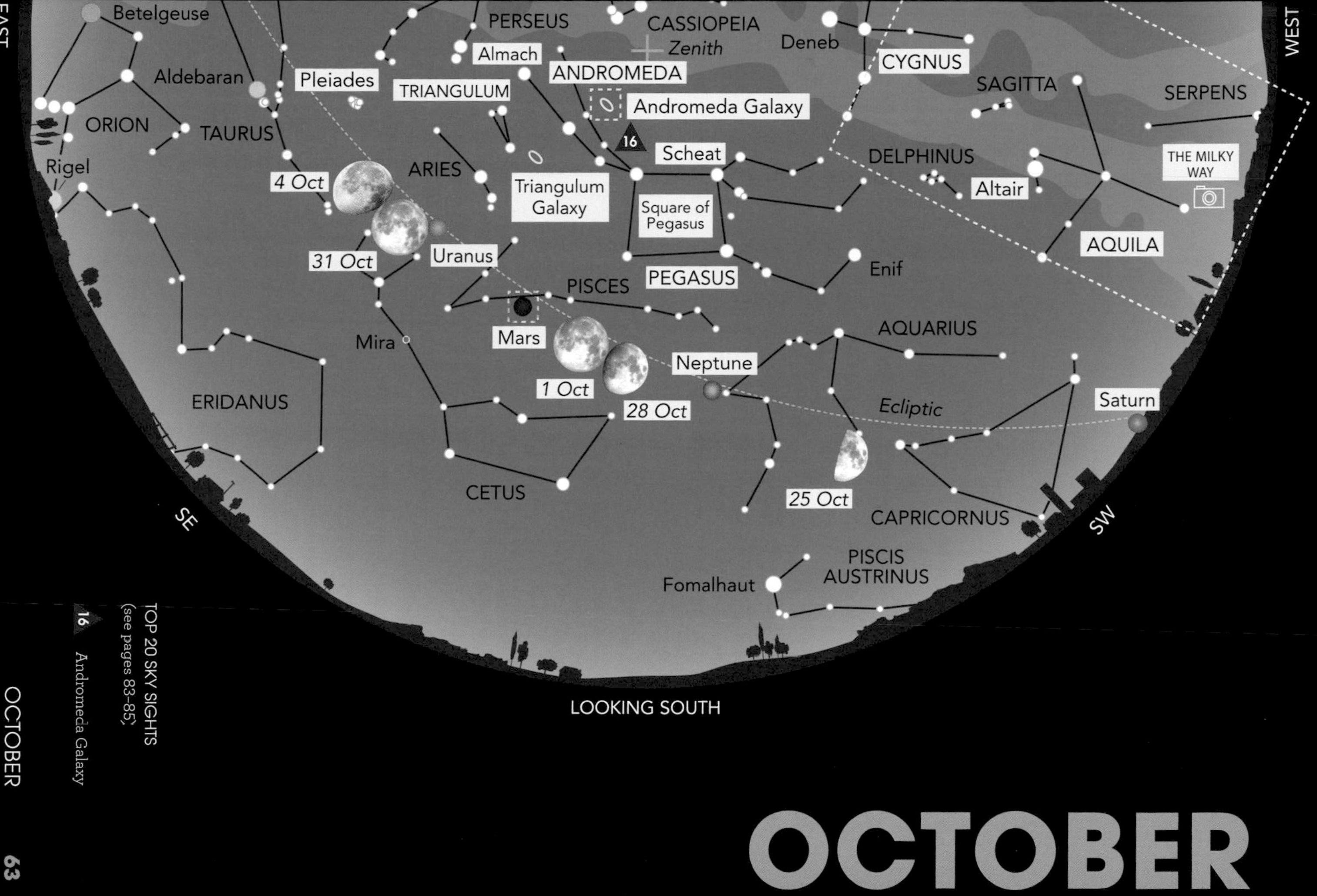

OCTOBER

TOP 20 SKY SIGHTS
(see pages 83–85)

16 Andromeda Galaxy

'Star' of the month is planet **Mars**, closer to the Earth this month than it's been for two years; brilliant in the evening sky, and the target of a bevy of spacecraft. Otherwise, the glories of October nights can best be described as 'subtle'. The barren **Square of Pegasus** dominates the southern sky, with **Andromeda** attached to his side. But the dull autumn constellations are already being faced down by the brighter lights of winter, spearheaded by the beautiful star cluster of the **Pleiades**.

OCTOBER'S CONSTELLATION

It takes considerable imagination to see the simple line of stars making up **Andromeda** as a young princess chained to a rock, about to be gobbled up by a sea monster – but that's ancient legends for you. Andromeda contains some surprising delights. **Almach**, the star at the left-hand end of the line, is a lovely sight in small telescopes: this beautiful double star comprises a yellow supergiant shining 2000 times brighter than the Sun, and a fainter bluish companion which is in fact triple.

But the glory of Andromeda is its great galaxy, beautifully placed on October nights. Lying above the line of stars, the **Andromeda Galaxy** is the most distant object easily visible to the unaided eye, a mind-boggling 2.5 million light years away. Containing 400 billion stars, the Andromeda Galaxy is the biggest member of the Local Group of galaxies. It is a wonderful sight in binoculars or a small telescope.

OBSERVING TIP

The Andromeda Galaxy is often described as the furthest object 'easily visible to the unaided eye'. It can be a bit elusive, though – especially if you are suffering from light pollution. The trick is to memorise Andromeda's pattern of stars, and then to look slightly to the *side* of where you expect the galaxy to be. This technique – called 'averted vision' – causes the image to fall on the outer region of your retina, which is more sensitive to light than the central region that's evolved to discern fine details. You'll certainly need averted vision to eyeball Andromeda's fainter sibling, the Triangulum Galaxy (see this month's Topic), which lies even further away. The technique is also crucial when you want to observe the faintest nebulae or galaxies through a telescope.

OCTOBER'S OBJECT

Mars – now at its closest to Earth – is the second-smallest planet in the Solar System, and the world most similar to our own. The Red Planet has polar caps, huge exposures of dark rocks, an atmosphere (very thin, and mainly carbon dioxide), seasons (twice as long as ours, because of Mars's distance from the Sun), and even clouds.

But what's most astonishing about Mars is its geology. All the extremes in the Solar System are here. The planet boasts an enormous canyon, Valles Marineris, which is 4000 kilometres long and seven kilometres deep. This enormous crack was caused by the upwelling of the Tharsis Ridge, home to family of vast volcanoes. Although not active today, it's just possible that they're dormant and could erupt tomorrow.

Largest of the volcanoes – and biggest in the Solar System – is Olympus Mons. With an altitude of 26 kilometres, the volcano is three times higher than Mount Everest. It would completely cover England, and its central crater could swallow London.

Five nations are using Mars' proximity to explore the planet with robotic spaceprobes in the next two years. Russia and the European Space Agency are developing ExoMars; NASA is building a 2020 rover; the Hope mission by Saudi Arabia is scheduled to be launched, as is the Chinese Mars mission, and India's Mars Orbiter 2 will join them. Expect a human being on Mars before 2040!

OCTOBER'S TOPIC: HOW FAR CAN YOU SEE?

A lot depends on light pollution (take a look at the maps on pages 92–95). If you live in a city, you're limited to just the brightest of stars – and the more luminous planets. Go to the suburbs, and you might see the Orion Nebula. It's a glowing star factory 1300 light years away.

Country dwellers have a plethora of sky-sights to gaze upon: star clusters, nebulae and even a galaxy far beyond our local star city. It's the Andromeda Galaxy (mentioned several times in this chapter), at 2.5 million light years away.

But – and this is probably down to keen eyesight – the most distant object that you can see with the unaided eye is the **Triangulum Galaxy** (close to the Andromeda Galaxy in the sky). A small, dim spiral galaxy, it lies three million light years away. All you need is a very dark desert to bring it into view.

OCTOBER'S PICTURE

A glorious image of the **Milky Way** – our spiral Galaxy viewed from within its disc – captured by Pete Williamson at Llyn Alwen (the Alwen Reservoir) near Bala in North Wales. Our Galaxy boasts magnificent spiral arms – but living in its flat suburbs, we never have the chance to marvel at them! It is home to 200,000 million stars – including our Sun – and has a massive black hole at its centre.

Pete Williamson took this picture in early October 2018 using a Canon 70D camera with a 14 mm f/2.8 lens. The single shot ISO 10,000 was captured at 25 seconds. He processed the image in Photoshop with Topaz noise removal.

OCTOBER'S CALENDAR

SUNDAY	MONDAY	TUESDAY	WEDNESDAY	THURSDAY	FRIDAY	SATURDAY
				1 10.05 pm Full Moon; Mercury E elongation	2 Moon near Mars	3 Moon near Mars (am); Venus very near Regulus
4	5 Moon near the Pleiades	6 Moon near Aldebaran and the Hyades	7	8	9	10 1.39 am Last Quarter Moon
11	12	13 Moon near Regulus (am); Mars opposition	14 Moon near Venus	15	16 8.31 pm New Moon	17
18	19	20	21 Orionids	22 Orionids (am); Moon near Jupiter	23 2.23 pm First Quarter Moon, near Saturn	24
25 BST ends	26	27	28	29 Moon near Mars	30	31 2.49 pm Full Moon, 'blue moon'; Uranus opposition

SPECIAL EVENTS

- **Night of 2/3 October:** the Moon is close to brilliant red Mars.
- **3 October:** Venus passes very close to Regulus in the morning (see Planet Watch).
- **13 October:** Mars lies opposite to the Sun in the sky, but due to its elliptical orbit the Red Planet is nearest to us on 6 October, at 62 million km.
- **14 October:** the crescent Moon forms a striking pair with brilliant Venus in the morning sky.
- **Night of 21/22 October:** maximum of the Orionid meteor shower, caused by debris from Halley's Comet smashing into Earth's atmosphere. Best observed after midnight, well after the Moon has set.
- **22 October:** the bright 'star' above the Moon is Jupiter, with Saturn to the left (Chart 10b).
- **23 October:** the Moon lies near Saturn and Jupiter.
- **25 October, 2 am:** the end of British Summer Time for this year, as clocks go back by an hour.
- **29 October:** the Moon passes under Mars.
- **31 October:** Uranus is opposite to the Sun in the sky and at its closest to Earth this year, 2811 million km away.
- **31 October:** the second Full Moon of the month is often called a 'blue moon' – nothing to do with its colour!
- **This month,** the European spacecraft BepiColombo swings past Venus on its way to Mercury.

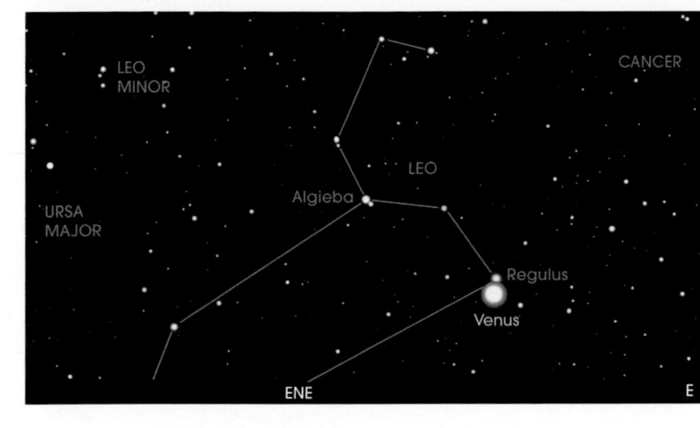

10a *3 October, 8 pm. Venus very close to Regulus.*

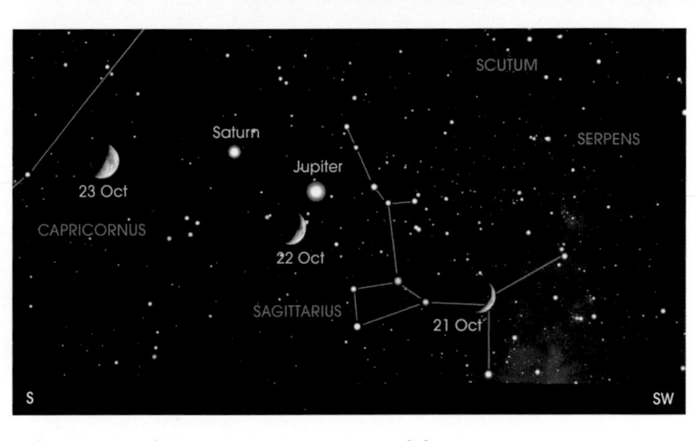

10b *21–23 October, 5 pm. Saturn, Jupiter and the Moon.*

• **Mars** is outshining everything else in the evening sky, nestled among the dim stars of Pisces. At opposition on 13 October, the Red Planet blazes at magnitude –2.6, and is visible all night long.

Mars

• Nearby **Uranus** also reaches peak magnitude at opposition this month – but 2000 times fainter at magnitude +5.7. With binoculars you can observe it throughout the night in Aries.

• Low in the south-west after sunset, brilliant **Jupiter** runs Mars a close second at magnitude –2.3. Lying in Sagittarius, the giant planet sets around 10.30 pm.

• **Saturn** lies to the left of Jupiter, also in Sagittarius. Fifteen times fainter at magnitude +0.5, the ringworld is setting about 11 pm.

• With binoculars or a telescope, you can spot **Neptune** (magnitude +7.8) in Aquarius, setting at around 4 am.

• Around 3.30 am, look out for resplendent **Venus** rising in the east: at magnitude –4.0, the Morning Star is even brighter than Mars. On the morning of 3 October (Chart 10a), Venus skims only 12 arcminutes from Regulus (which is 100 times fainter): at first glance, it looks as though Regulus has gone supernova!

• **Mercury** is lost in the Sun's glare during October.

OCTOBER'S PLANET WATCH

- The sky at 10 pm in mid-November, with Moon positions at three-day intervals either side of Full Moon.
- The star positions are also correct for 11 pm at the beginning of November, and 9 pm at the end of the month.
- The planets move slightly relative to the stars during the month.

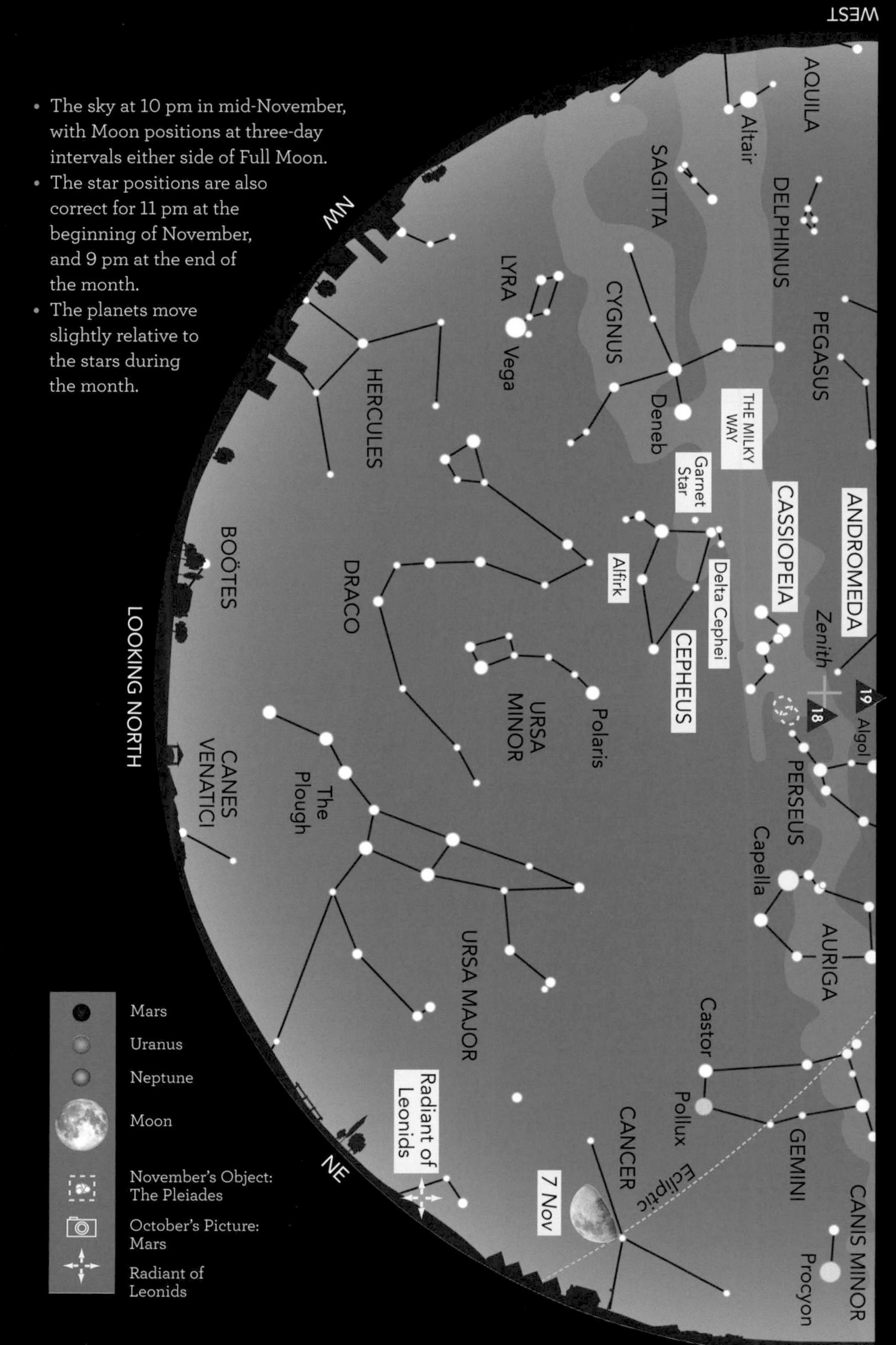

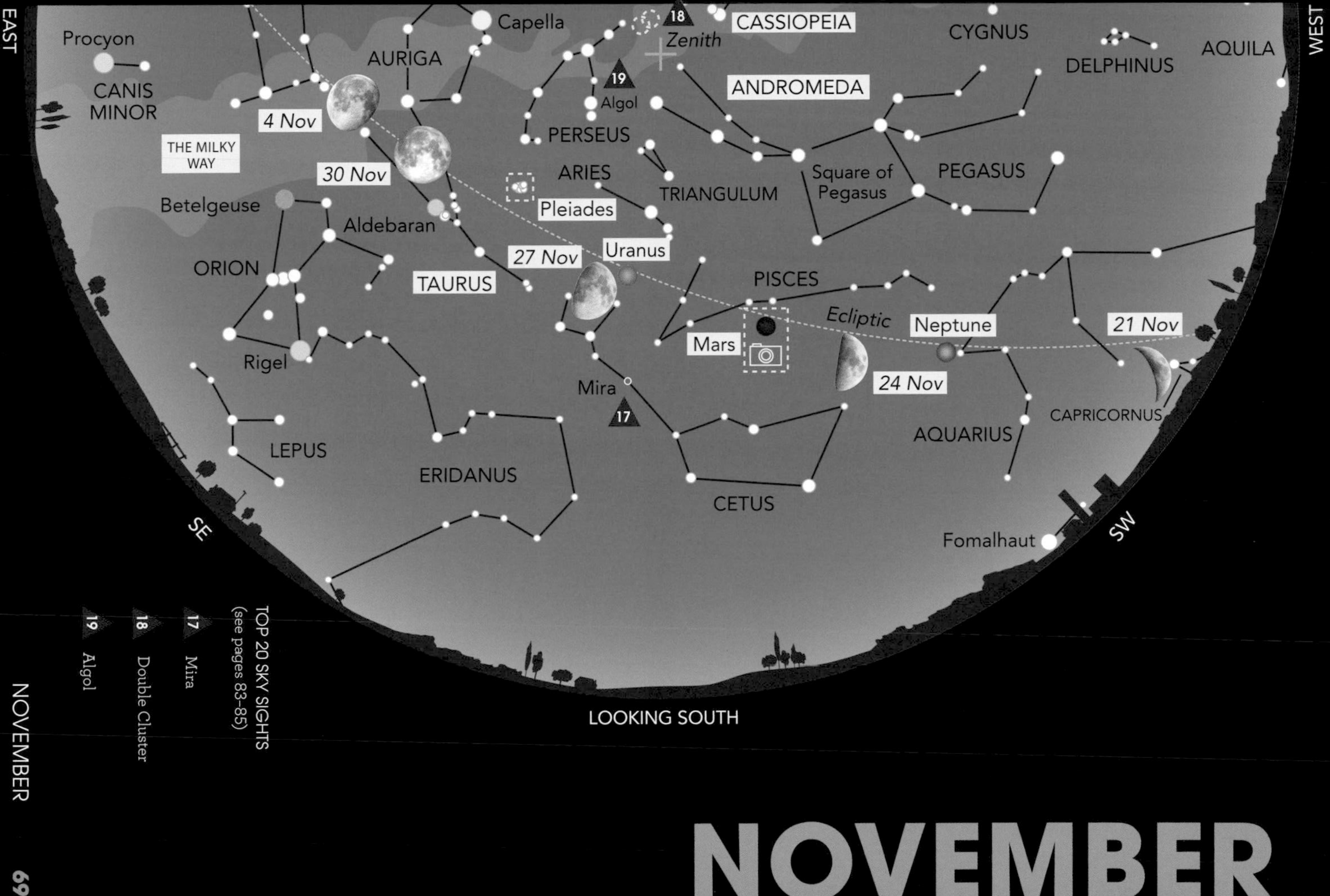

NOVEMBER

TOP 20 SKY SIGHTS
(see pages 83–85)

17 Mira

18 Double Cluster

19 Algol

The **Milky Way** rears overhead on these dark November nights, providing a stunning inside perspective on the huge Galaxy that is our home in space. Look carefully, and you can see that it's spangled with fuzzy glowing diadems. Even better, sweep the band of the Milky Way with binoculars or a small telescope, and these blurry jewels appear in their true guise: distant clusters of stars.

NOVEMBER'S CONSTELLATION

The triangular constellation of **Cepheus** is meant to represent the King of Ethiopia, married to the magnificent next-door constellation **Cassiopeia**. Both in legend and visually, his wife is far more exciting (she once boasted that her daughter **Andromeda** was more beautiful than all the sea nymphs, with almost disastrous effects). As a constellation, Cepheus is faint and somewhat boring – save for a trio of fascinating stars. **Alfirk** is a double star, with the companion being visible through a small telescope. The aptly named hypergiant **Garnet Star** – named by William Herschel because of its strong ruddy hue – changes in brightness between magnitudes +3.4 and +5.1 with an approximate period of two years. But Cepheus is home to the most iconic of all variable stars – **Delta Cephei**. This star changes in brightness (from magnitude +3.5 to +4.4) over a period of five days and nine hours. Astronomers have discovered that this particular class of star (Cepheids) has variation timescales related to their intrinsic luminosities – allowing them to be used as pulsating stellar beacons to measure cosmic distances.

OBSERVING TIP

With Christmas on the way, you may well be thinking of buying a telescope as a present for a budding stargazer. Beware! Unscrupulous websites and mail-order catalogues often advertise small telescopes that boast huge magnifications. This is 'empty magnification' – blowing up an image that the lens or mirror simply doesn't have the ability to get to grips with, so all you see is a bigger blur. The maximum magnification a telescope can actually provide is twice the diameter of the lens or mirror in millimetres. So if you see an advertisement for a 75 mm telescope, beware of any claims for a magnification greater than 150 times.

NOVEMBER'S OBJECT

The **Pleiades** star cluster is one of the most familiar sky-sights. It is lovely seen with the naked eye or through binoculars, and magnificent in a long-exposure image.

Though the cluster is well known as the Seven Sisters, skywatchers typically see any number of stars but seven! Most people can pick out the six brightest stars, while very keen-sighted observers can discern up to 11 members. These are just the most luminous in a group of at least 1000 stars, lying 444 light years away. The brightest stars in the Pleiades are hot and blue, and all the stars are young – around 100 million years old.

The space observatory Spitzer has detected a disc around one of the Pleiades' stars. The hot money is that it's a planetary system in formation.

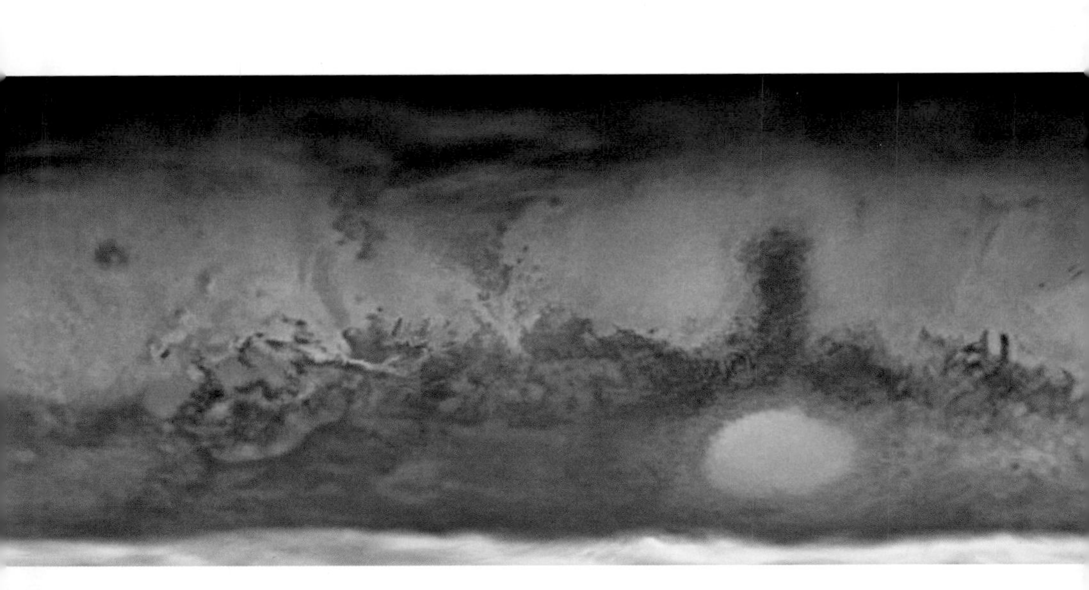

Damian Peach used the 1 m Chilescope in South America to map the Red Planet. You can use the telescope remotely, but Damian was on-site in August 2018 for most of the images, which he captured with a ZWO ASI174MM camera. He used about ten separate images to cover the whole planet.

NOVEMBER'S TOPIC: WILLIAM HERSCHEL

Herschel – a German composer and musician, based in Bath – has the distinction of being the first person in history to discover a new planet in the Solar System. Music was his career, but astronomy was his passion. He revelled in building bigger and bigger telescopes. One night in 1781, he was scanning the skies from his back garden; his project was to work out the geography of the Universe.

He came across a greenish blob he'd never seen before. The next night, it had moved. Herschel thought it was a comet; but scientists at the Royal Society in London realised that it was moving too slowly. It had to be a planet.

But what to call it? Herschel wanted to name the planet 'Georgium Sidus', after King George III (the only monarch known to have an interest in science). But tradition prevailed, and it became Uranus – the father of Saturn.

Discovering Uranus doubled the size of the Solar System overnight. King George honoured Herschel by installing him in a home close to Windsor Castle, where he was the astronomer to the king. It was there that Herschel mapped out the lens-shape of our Galaxy.

NOVEMBER'S PICTURE

The renowned British astrophotographer Damian Peach constructed this Mercator-projection map of Mars using a telescope in Chile. The south polar cap lies along the bottom; above and to the right, is Hellas – a huge, bright impact basin. Projecting above Hellas is Syrtis Major, a tongue of dark rocks littering the Martian desert; it's easily visible through medium-sized telescopes. Far left – in the desert – is a circular blob. This is Olympus Mons, the tallest volcano in the Solar System.

NOVEMBER'S CALENDAR

SUNDAY	MONDAY	TUESDAY	WEDNESDAY	THURSDAY	FRIDAY	SATURDAY
1 Moon near the Pleiades	2 Moon between the Hyades and Pleiades, near Aldebaran	3	4	5	6 Moon near Castor and Pollux	7 Moon near Praesepe
8 1.46 pm Last Quarter Moon	9 Moon near Regulus (am)	10 Mercury W elongation	11	12 Moon above Venus (am)	13 Moon between Venus and Mercury (am)	14
15 5.07 am New Moon; Venus near Spica (am)	16 Venus near Spica (am)	17 Leonid meteor shower	18 Leonid meteor shower (am)	19 Moon near Jupiter and Saturn	20	21
22 4.45 am First Quarter Moon	23	24	25 Moon near Mars	26	27	28
29 Moon between the Hyades and Pleiades, near Aldebaran	30 9.29 am Full Moon, near Aldebaran					

SPECIAL EVENTS

- **12 November:** a lovely sight this morning, as the crescent Moon lies directly above brilliant Venus (Chart 11a).
- **13 November:** look to the left of Venus before dawn, to spot a very narrow crescent Moon: Mercury lies to the lower left again (Chart 11a).

Leonid meteor shower

- **Night of 17/18 November:** maximum of the **Leonid meteor shower**. Occasionally the Leonids have stormed the Earth with shooting stars; but we're not expecting any great fireworks this year as its parent body, Comet Tempel-Tuttle, is far away from us.
- **19 November:** the bright 'star' to the right of the crescent Moon is Jupiter, with Saturn above and between them (Chart 11b).
- **25 November:** Mars lies above the Moon.

NOVEMBER'S PLANET WATCH

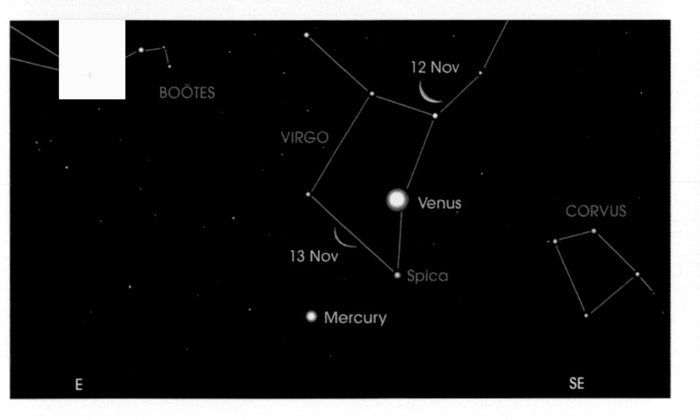

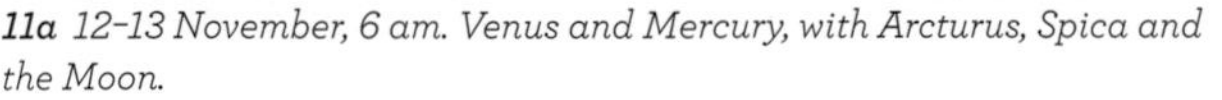

11a *12–13 November, 6 am. Venus and Mercury, with Arcturus, Spica and the Moon.*

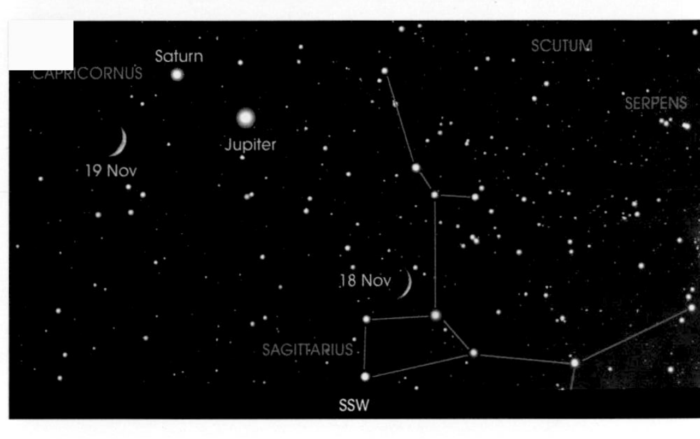

11b *18–19 November, 6 pm. Saturn, Jupiter and the crescent Moon.*

- The brilliant 'star' you'll see on the south-west after sunset is **Jupiter**, shining at magnitude –2.1 in Sagittarius, and setting about 8 pm. Immediately to its left is rather fainter **Saturn**, also in Sagittarius. At magnitude +0.6, the ringworld sinks beneath the horizon 20 minutes after Jupiter. During the month, watch how the planets are steadily approaching one another.
- In the south-east, **Mars** is rapidly fading after its close encounter with Earth last month, dropping from magnitude –2.0 to –1.1 by the end of November. Residing in Pisces, the Red Planet sets around 3.30 am.
- **Neptune** (magnitude +7.8) lies in Aquarius, and sets about 1 am. Its near-twin **Uranus** is in Pisces: at magnitude +5.7, it is setting around 6 am.
- Glorious **Venus** is rising in the east at around 4 am, and shining at magnitude –4.0. On the mornings of 15 and 16 November it passes near Spica.
- **Mercury** is putting on its best morning performance of the year in the middle of November, rising above the eastern horizon around 5.30 am. Seek out the elusive planet to the lower left of Venus, from roughly 5 November when it's at magnitude +0.1 and lying to the left of rather dimmer Spica. Mercury continues brightening to magnitude –0.7 by 25 November, when it slips down into the dawn twilight.

- The sky at 10 pm in mid-December, with Moon positions at three-day intervals either side of Full Moon.
- The star positions are also correct for 11 pm at the beginning of December, and 9 pm at the end of the month.
- The planets move slightly relative to the stars during the month.

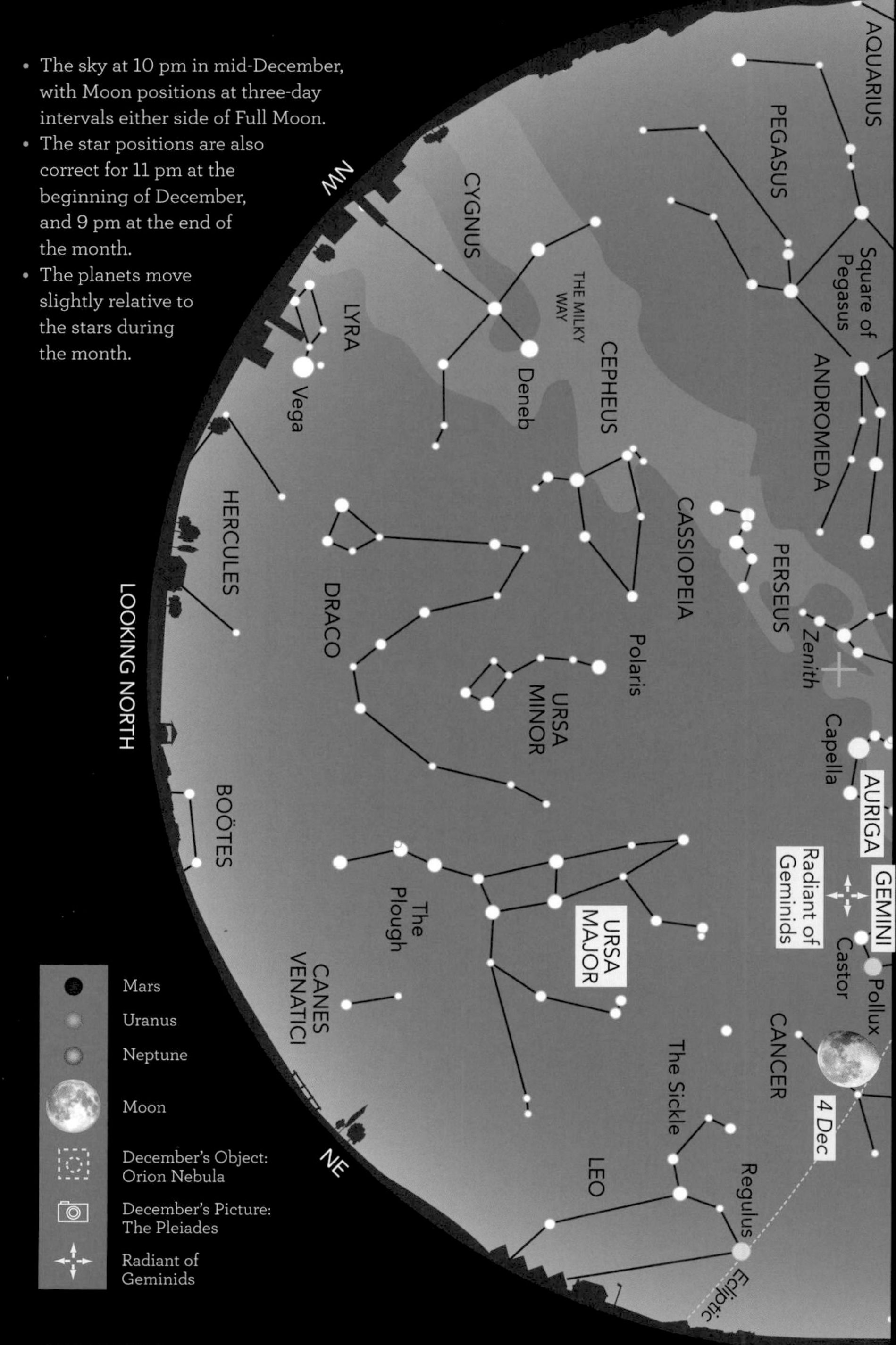

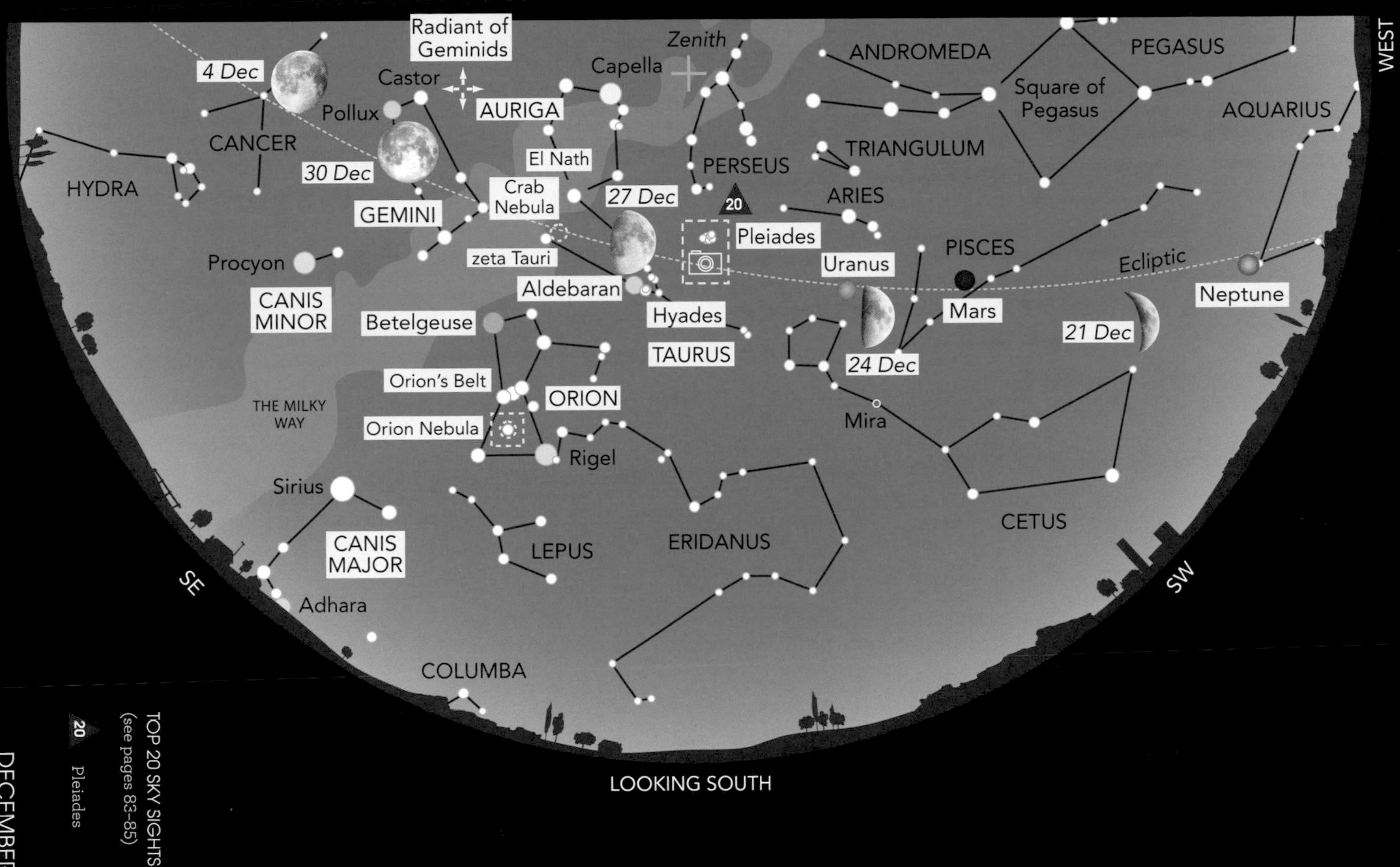

TOP 20 SKY SIGHTS
(see pages 83–85)

20 Pleiades

You'll find the Solar System's two giants really up close and personal low on the horizon: on 21 December, Jupiter and Saturn pass closer than at any time since July 1623! We're also treated to the best shooting-star display of the year, and the regular brilliant constellations of winter: **Orion**, his hunting dogs **Canis Major** and **Canis Minor**, **Taurus** (the Bull), the hero twins of **Gemini**, and the charioteer **Auriga** almost overhead.

DECEMBER'S CONSTELLATION

Taurus is very much a second cousin to brilliant **Orion**, but a fascinating constellation nonetheless. It's dominated by **Aldebaran**, the baleful blood-red eye of the celestial bull. Around 65 light years away, and shining with a (slightly variable) magnitude of +0.85, Aldebaran is a red giant star, a tad more massive than the Sun.

The 'head' of the bull is formed by the **Hyades** star cluster (see January's Picture). The other famous star cluster in Taurus is the far more glamorous **Pleiades** (see November's Object), whose stars – although further away than the Hyades – are younger and brighter.

Taurus has two 'horns': the star **El Nath** (Arabic for 'the butting one') to the north, and **zeta Tauri** (whose Babylonian name Shurnarkabti-sha-shutu – meaning 'star in the bull towards the south' – is thankfully not generally used!). Above this star is a stellar wreck – literally. In 1054, Chinese astronomers witnessed a brilliant 'new star' appear in this spot, visible in daytime for weeks. It was a supernova – an exploding star in its death throes. And today, through a medium-sized telescope, we see the still-expanding remains as the **Crab Nebula**.

OBSERVING TIP

Hold a 'meteor party' to check out the year's best celestial firework show, the Geminid meteor shower on 13/14 December. You don't need any optical equipment – in fact, telescopes and binoculars will restrict your view of the shooting stars, which can appear anywhere. The ideal viewing equipment is your unaided eye, plus a warm sleeping bag and a lounger. Everyone should look in different directions, so you can cover the whole sky: shout out 'Meteor!' when you see a shooting star. One of the party can record the observations, using a watch, notepad and red torch. In the interests of science, try to brave the cold and observe the sky for at least an hour, before repairing indoors for some warming seasonal cheer . . .

DECEMBER'S OBJECT

Below the three stars of **Orion's Belt** lies a fuzzy patch. Through binoculars or a telescope it looks like a small glowing cloud in space. It *is* a cloud – but at 24 light years across, it's hardly petite. Only the distance of the **Orion Nebula** – 1300 light years – diminishes it. It's part of a vast region of starbirth in Orion, and the nearest region to Earth where heavyweight stars are being born. This star factory contains at least 700 fledgling stars, which have just hatched out of immense dark clouds of dust and gas: most are visible only with a telescope that picks up infrared (heat) radiation.

DECEMBER'S TOPIC: BLACK HOLES

A black hole is the densest concentration of matter in the Universe, and the most mysterious. For starters, light (and all other radiation) can't escape its gravity – so, by definition, it's black. And it's a hole. Nothing can escape its clutches: whatever falls in is trapped forever, because nothing can travel faster than light.

When a supermassive star explodes at the end of its life, its core can implode and collapse into a black hole. The resulting cosmic abyss is tiny; no more than 30 kilometres across.

So how do you find black holes in our mighty cosmos? Well . . . imagine a black cat having an altercation with a white cat in a dark cellar. You won't see the black cat. Watch the white cat being swung around, though, and you'll realise there's a dark presence, too – and you can even hazard a guess as to its weight.

Jamie Cooper used a Canon 5D Mark III camera with a 200 mm f/2.8L lens to take this image. He used an exposure of 30 seconds, at ISO 3200, stacking five exposures together to reduce image noise, and finessed the composition in Photoshop.

And so it is with black holes. When one is in orbit around a normal star ('the white cat'), it drags streams of gas off its surface, and this material ends up in a superheated accretion disc circling the hole. The dazzling radiation from this disc is the giveaway.

And what's the fate of the matter that falls into a black hole? Some scientists have calculated that it enters another universe: meaning that black holes could be gateways to a new cosmos.

DECEMBER'S PICTURE

Jamie Cooper captured this portrait of Comet Wirtanen next to the **Pleiades** star cluster on 15 December 2018 – when the comet was at its closest to the Earth (12 million kilometres, or 30 times the Moon's distance). The small comet orbits the Sun every 5.4 years, and – on this unusually close encounter – astronomers hoped it would put on a good display. Slightly dimmer than expected, Wirtanen was *just* visible to the unaided eye in the countryside as a fuzzy blob. Perspective made its tail slink behind it, and all that was visible was the gaseous coma: heated ices boiling off the comet.

DECEMBER'S CALENDAR

SUNDAY	MONDAY	TUESDAY	WEDNESDAY	THURSDAY	FRIDAY	SATURDAY
		1	2	3 Moon near Castor and Pollux	4 Moon near Praesepe	5
6 Moon near Regulus	7	8 0.36 am Last Quarter Moon	9	10 Moon near Spica (am)	11 Moon near Spica (am)	12 Moon near Venus (am)
13 Geminids	14 Geminids (am); 4.16 pm New Moon; solar eclipse	15	16	17 Moon near Jupiter and Saturn	18	19
20	21 11.41 pm First Quarter Moon; Winter Solstice; Jupiter extremely close to Saturn	22	23 Moon near Mars	24 Venus near Antares (am)	25	26 Moon near the Pleiades
27 Moon between the Hyades and Pleiades, near Aldebaran	28	29	30 3.28 am Full Moon, near Castor and Pollux	31 Moon near Castor and Pollux		

SPECIAL EVENTS

- **12 December:** the crescent Moon hangs to the upper right of Venus in the morning sky (Chart 12a).
- **Night of 13/14 December:** look out for the bright slow shooting stars of the **Geminid meteor shower**, which – unusually – are debris not from a comet, but from an asteroid, called Phaethon. It's a perfect year for observing this prolific display, as the Moon is well out of the way.
- **14 December:** a total eclipse of the Sun is visible from a narrow track starting in the South Pacific, crossing southern Chile and Argentina, and ending in the South Atlantic: the maximum duration is 2 minutes 9.6 seconds. Much of South America experiences a partial eclipse, but nothing is visible from the British Isles.
- **17 December:** the thinnest crescent Moon lies just to the left of the close pairing of Jupiter and Saturn (see Planet Watch).
- **21 December, 10.02 am:** the Winter Solstice, when the Sun reaches its lowest point in the sky as seen from the northern hemisphere, giving us the shortest day and longest night.
- **21 December:** Jupiter and Saturn are incredibly close together (see Planet Watch).
- **23 December:** Mars lies above the Moon.

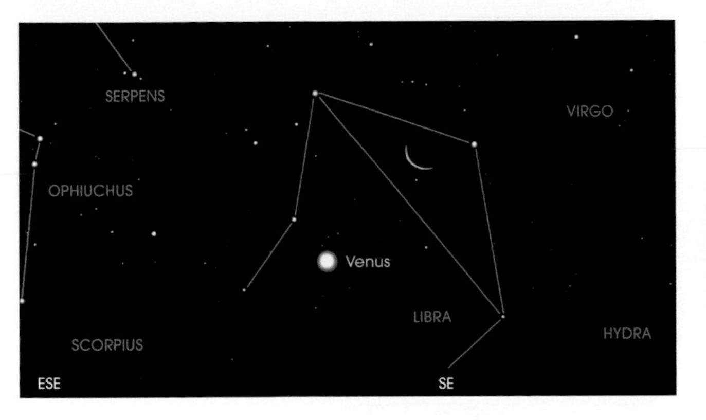

12a *12 December, 6.30 pm. Venus with the crescent Moon.*

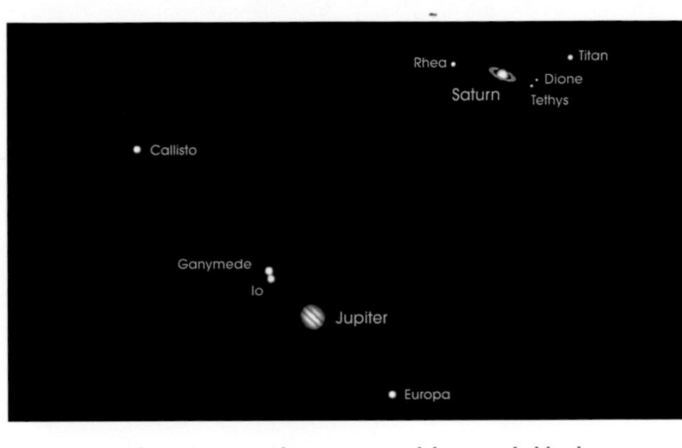

12b *21 December, 5.30 pm. Telescopic view of the remarkably close encounter of Jupiter and Saturn, showing the planets and their brightest moons.*

- Very low in the south-west, keep an eye on the giant planets **Jupiter** (magnitude –2.0) and **Saturn** (10 times fainter at magnitude +0.6) before they set about 6.30 pm. During December, the two planets (both in Sagittarius) creep ever nearer to one another. On 21 December they are just 6 arcminutes apart: to the naked eye they'll almost seem to merge together. Though they'll be right on the horizon, grab a telescope if you can for this unique opportunity to see the stripy disc of Jupiter and Saturn's rings in the same field of view (Chart 12b).

Jupiter

- **Mars** is brilliant in the southern sky, though fading from magnitude –1.1 to –0.2 during December. You'll find the Red Planet in Pisces, setting about 2.30 am.
- Lying in Aquarius, **Neptune** is below naked-eye visibility at magnitude +7.9 and sets around 11 pm. Setting about 4 am, **Uranus** (magnitude +5.7) lies in Aries.
- **Venus** is a glorious Morning Star, rising in the south-east around 5.30 am. On the morning of 24 December, Venus passes above Antares.
- **Mercury** is too close to the Sun to be seen this month.

DECEMBER'S PLANET WATCH

Can you see the planets? We're amazed when people ask us that question: some of our cosmic neighbours are the brightest objects in the night sky after the Moon. As they're so close, you can watch them getting up to their antics from night to night. And planetary debris – leftovers from the birth of the Solar System – can light up our skies as glowing comets and the celestial fireworks of a meteor shower.

THE SUN-HUGGERS

Mercury and Venus orbit the Sun more closely than our own planet, so they never seem to stray far from our local star: you can spot them in the west after sunset, or the east before dawn, but never all night long. At *elongation*, the planet is at its greatest separation from the Sun, though – as you can see in the diagram (right) – that's not when the planet is at its brightest. Through a telescope, Mercury and Venus (technically known as the *inferior planets*) show phases like the Moon – from a thin crescent to a full globe – as they orbit the Sun.

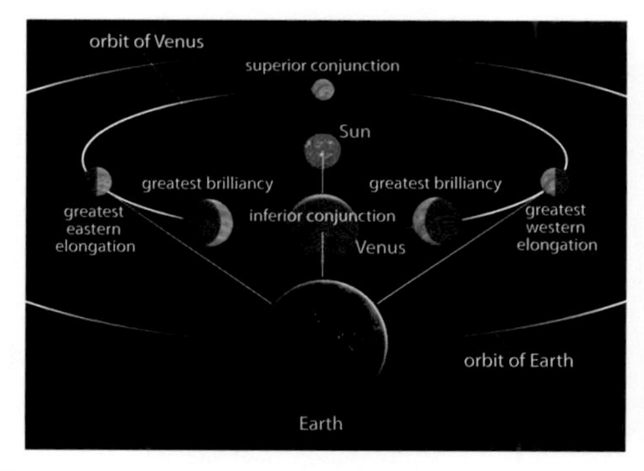

Venus (and Mercury) show phases like the Moon as they orbit the Sun.

Mercury

The innermost planet makes its best evening appearance in February, and reappears in the dusk sky during May–June – though its evening appearance in October is lost in bright twilight. Mercury is difficult to spot in the dawn twilight at its March apparition, but the elusive planet puts on a better morning show in July–August. It's best seen before dawn in November.

Maximum elongations of Mercury in 2020	
Date	**Separation**
10 February	18° east
24 March	28° west
4 June	24° east
22 July	20° west
1 October	26° east
10 November	19° west

Venus

In early 2020, you'd be forgiven for thinking brilliant Venus is a permanent fixture in the evening sky, hanging in the west for hours after sunset and reaching maximum brightness on **28 April**. But at the end of May, the Evening Star dives down to the horizon to reappear in the morning sky in June. Venus then hangs around as the Morning Star until the end of the year.

Maximum elongations of Venus in 2020	
Date	**Separation**
24 March	46° east
13 August	46° west

WORLDS BEYOND

A planet orbiting the Sun beyond the Earth (known in the jargon as a *superior planet*) is visible at all times of night, as we look outwards into the Solar System. It lies due south at midnight when the Sun, the Earth and the planet are all in line – a time known as *opposition* (see the diagram, right). Around this time the Earth lies nearest to the planet, although the date of closest approach (and the planet's maximum brightness) may differ by a few days because the planets' orbits are not circular.

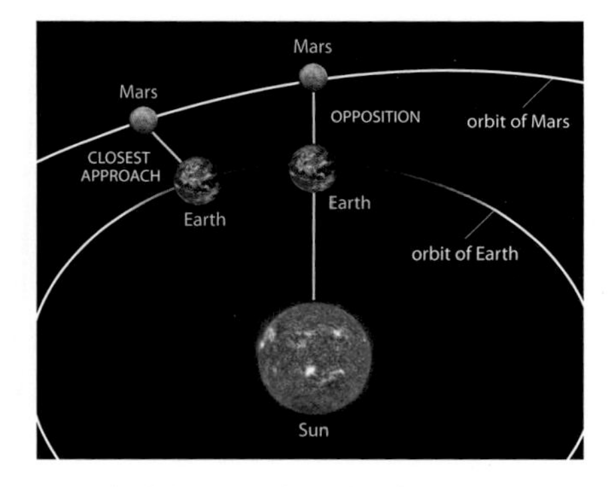

Mars (and the outer planets) line up with the Sun and Earth at opposition, but they are brightest at their point of closest approach.

Mars

For the first half of the year, you'll need to be up after midnight to catch the Red Planet. In late summer, Mars brightens rapidly towards closest approach on **6 October** and opposition on **13 October**, when it outshines everything (bar the Moon) in the evening sky. For the rest of the year, it gradually fades.

● Where to find Mars	
Early January	Libra
Mid-January	Scorpius
Late January to mid-February	Ophiuchus
Mid-February to March	Sagittarius
April	Capricornus
May to June	Aquarius
July to December	Pisces

Jupiter

The giant planet comes into view in the morning sky during February; by June it's rising before midnight. Jupiter reaches opposition on **14 July**, and is visible in the evening sky until the end of the year. Throughout 2020, Jupiter lies in Sagittarius.

Saturn

The ringed planet lies near Jupiter the whole year, culminating in a historic close encounter on 21 December. Appearing in the morning sky in late February, Saturn moves from Sagittarius into Capricornus, and then back to Sagittarius in July where it remains for the rest of the year. Saturn is at opposition on **20 July**.

Uranus

Just perceptible to the naked eye, Uranus lies in Aries all year. Up until March, the seventh planet is visible in the evening sky. It emerges from the Sun's glow in the morning sky in June. Uranus is at opposition on **31 October**.

Neptune

The most distant planet lies in Aquarius throughout the year, and is at opposition on **11 September**. Neptune can be seen (though only through binoculars or a telescope) in January and February – watch out for a close encounter with Venus on 27 January – and then from May until the end of the year.

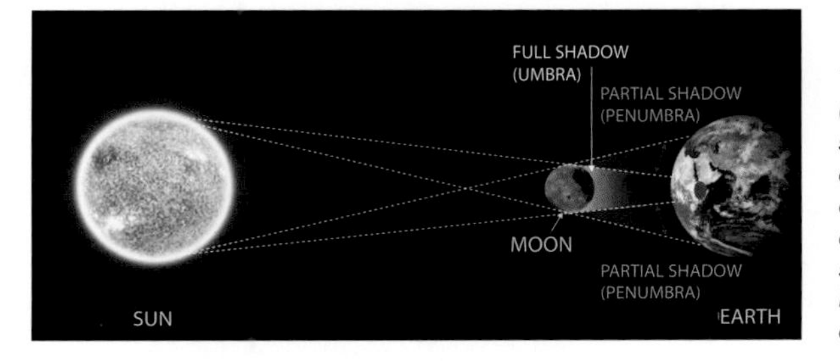

Where the dark central part (the umbra) of the Moon's shadow reaches the Earth, we are treated to a total solar eclipse. If the shadow doesn't quite reach the ground, we see an annular eclipse. People located within the penumbra observe a partial eclipse.

SOLAR ECLIPSES

An annular eclipse (when we see a thin ring of the Sun's surface around the Moon's silhouette) is visible on **21 June** from a narrow band crossing central Africa, northern India and China. From the British Isles, no eclipse will be seen.

On **14 December**, a total eclipse of the Sun is visible from a narrow track starting in the South Pacific, crossing southern Chile and Argentina, and ending in the South Atlantic; it's not visible from the British Isles.

LUNAR ECLIPSES

There are no total, or even partial, eclipses of the Moon in 2020. The best we can see from the British Isles is a penumbral eclipse on **10 January**. The lower edge of the Full Moon looks slightly tarnished as some – but not all – of the sunlight falling on the Moon is blocked by the Earth.

METEOR SHOWERS

Shooting stars, or *meteors,* are tiny specks of interplanetary dust, burning up in the Earth's atmosphere. At certain times of year, Earth passes through a stream of debris (usually left by a comet) and we see a *meteor shower*. The meteors appear to emanate from a point in the sky known as the *radiant*. Most showers are known by the constellation in which the radiant lies.

Table of meteor showers	
Meteor shower	**Date of maximum**
Quadrantids	3/4 January
Lyrids	21/22 April
Eta Aquarids	5/6 May
Perseids	12/13 August
Orionids	21/22 October
Leonids	17/18 November
Geminids	13/14 December

It's fun and rewarding to hold a meteor party. Note the location, cloud cover, the time and brightness of each meteor and its direction through the stars – along with any persistent afterglow (train).

COMETS

Comets are dirty snowballs from the outer Solar System. If they fall towards the Sun, its heat evaporates their ices to produce a gaseous head (*coma*) and sometimes dramatic tails. Although some comets are visible to the naked eye, you'll need a telescope to see the fine detail in the coma.

Hundreds of comets move round the Sun in small orbits. But many more don't return for thousands or even millions of years. Most comets are now discovered in professional surveys of the sky, but a few are still found by dedicated amateur astronomers. And watch out in case a brilliant new comet puts in a surprise appearance!

We've always had our favourite sights in the night sky: and here they are in a season-by-season summary. It doesn't matter if you're a complete beginner, finding your way around the heavens with the unaided eye or binoculars; or if you're a seasoned stargazer, with a moderate telescope. There's something here for everyone.

Each sky-sight comes with a brief description, and a guide as to how you can best see it. Many of the most delectable objects are faint, so avoid moonlight when you go out spotting. Most of all, enjoy!

SPRING

Praesepe

Constellation: Cancer
Star Chart/Key: March; 5
Type/Distance: Star cluster; 600 light years
Magnitude: +3.7
A fuzzy patch to the unaided eye; a telescope reveals many of its 1000 stars.

M81 and M82

Constellation: Ursa Major
Star Chart/Key: March; 6
Type/Distance: Galaxies; 12 million light years
Magnitude: +6.9 (M81); +8.4 (M82)
A pair of interacting galaxies: the spiral M81 appears as an oval blur, and the starburst M82 as a streak of light.

The Plough

Constellation: Ursa Major
Star Chart/Key: April; 7
Type/Distance: Asterism; 82–123 light years
Magnitude: Stars are roughly magnitude +2
The seven brightest stars of the Great Bear form a large saucepan shape, called 'the Plough'.

Virgo Cluster

Mizar and Alcor

Constellation: Ursa Major
Star Chart/Key: April; 8
Type/Distance: Double star; 80 light years
Magnitude: +2.3 (Mizar); +4.0 (Alcor)
The sky's classic double star, easily separated by the unaided eye: a telescope reveals Mizar itself is a close double.

Virgo Cluster

(difficult)

Constellation: Virgo
Star Chart/Key: May; 9
Type/Distance: Galaxy cluster; 54 million light years
Magnitude: Galaxies range from magnitude +9.4 downwards
Huge cluster of 2000 galaxies, best seen through moderate to large telescopes.

SUMMER

Antares

Constellation: Scorpius
Star Chart/Key: June; 10
Type/Distance: Double star; 550 light years
Magnitude: +0.96
Bright red star close to the horizon. You can spot a faint green companion with a telescope.

M13

Constellation: Hercules
Star Chart/Key: June; 11
Type/Distance: Star cluster; 22,200 light years
Magnitude: +5.8

Dumbbell Nebula

A faint blur to the naked eye, this ancient globular cluster is a delight seen through binoculars or a telescope. It boasts nearly a million stars.

Lagoon and Trifid Nebulae

Constellation: Sagittarius
Star Chart/Key: July; 12
Type/Distance: Nebulae; 4000 light years (Lagoon); 5200 (Trifid)
Magnitude: +6.0 (Lagoon); +7.0 (Trifid)
While the Lagoon Nebula is just visible to the unaided eye, you'll need binoculars or a telescope to spot the Trifid. The two are in the same telescope field of view, and present a stunning photo opportunity.

Albireo

Constellation: Cygnus
Star Chart/Key: August; 13
Type/Distance: Double star; 415 light years
Magnitude: Albireo A: +3.2; Albireo B: +5.1
Good binoculars reveal Albireo as being double. But you'll need a small telescope to appreciate its full glory. The brighter star appears golden; its companion shines piercing sapphire. It is the most beautiful double star in the sky.

Dumbbell Nebula

Constellation: Vulpecula
Star Chart/Key: August; 14
Type/Distance: Planetary nebula; 1300 light years
Magnitude: +7.5
Visible through binoculars, and a lovely sight through a small/medium telescope, this is a dying star that has puffed off its atmosphere into space.

AUTUMN

Delta Cephei

Constellation: Cepheus
Star Chart/Key: September; 15
Type/Distance: Variable star; 890 light years
Magnitude: +3.5 to +4.4, varying over 5 days 9 hours
The classic variable star, Delta Cephei is chief of the Cepheids – stars that allow us to measure distances in the Universe (their variability time is coupled to their intrinsic luminosity). Visible to the unaided eye, but you'll need binoculars for serious observations.

Andromeda Galaxy

Constellation: Andromeda
Star Chart/Key: October; 16
Type/Distance: Galaxy; 2.5 million light years
Magnitude: +3.4
The biggest major galaxy to our own, the Andromeda Galaxy is easily visible to the unaided eye in unpolluted skies. Four times the width of the Full Moon, it's a great telescopic object and photographic target.

Double Cluster

Mira

Constellation: Cetus
Star Chart/Key: November; 17
Type/Distance: Variable star; 300 light years
Magnitude: +3.5 to +10.1 over 332 days, although maxima and minima may vary
Mira is not nicknamed 'the Wonderful' for nothing. This distended red giant star is

alarmingly variable as it swells and shrinks At its brightest, it's a naked-eye object; binoculars may catch it at minimum; but you need a telescope to monitor this star. Its behaviour is highly unpredictable, and it's important to keep logging it.

Double Cluster

Constellation: Perseus
Star Chart/Key: November; 18
Type/Distance: Star clusters; 7500 light years
Magnitude: +3.7 and +3.8
A lovely sight to the unaided eye, these stunning young star clusters are sensational through binoculars or a small telescope. They're a great photographic target.

Algol

Constellation: Perseus
Star Chart/Key: November; 19
Type/Distance: Variable star; 90 light years
Magnitude: +2.1 to +3.4 over 2 days 21 hours
Like Mira, Algol is a variable star, but not an intrinsic one. It's an 'eclipsing binary' – its brightness falls when a fainter companion star periodically passes in front of the main star. It's easily monitored by the eye, binoculars or a telescope.

WINTER

Pleiades

Constellation: Taurus
Star Chart/Key: December; 20
Type/Distance: Star cluster; 444 light years
Magnitude: Stars range from magnitude +2.9 downwards
To the naked eye, most people can see six stars in the cluster, but it can rise to 14 for the keen-sighted. In binoculars or a telescope, they are a must-see. Astronomers have observed 1000 stars in the Pleiades.

Orion Nebula

Constellation: Orion
Star Chart/Key: January; 1
Type/Distance: Nebula; 1300 light years
Magnitude: +4.0
A striking sight even to the unaided eye, the

Pleiades

Orion Nebula – a star-forming region 24 light years across – hangs just below Orion's belt. Through binoculars or a small telescope, it is staggering. A photographic must!

Betelgeuse

Constellation: Orion
Star Chart/Key: January; 2
Type/Distance: Variable star; 720 light years
Magnitude: 0.0 – +1.3
Even with the unaided eye, you can see that Betelgeuse is slightly variable over months, as the red giant star billows in and out.

M35

Constellation: Gemini
Star Chart/Key: February; 3
Type/Distance: Star cluster; 2800 light years
Magnitude: +5.3
Just visible to the unaided eye, this cluster of around 2000 stars is a lovely sight through a small telescope.

Sirius

Constellation: Canis Major
Star Chart/Key: February; 4
Type/Distance: Double star; 8.6 light years
Magnitude: –1.47
You can't miss the Dog Star. It's the brightest star in the sky! But you'll need a 150 mm reflecting telescope (preferably bigger) to pick out its +8.44 magnitude companion – a white dwarf nicknamed 'the Pup'.

These days photography is part of our lives as never before. Many of us carry phones that will take a brilliant photo, and we use these to document our lives far more than when we needed to carry a separate camera. Inevitably we want to be able to photograph the sights that we see through our telescopes. The photos taken by UK amateurs in *Stargazing 2020* are a great encouragement. So how does one go about it?

Many phones or cameras can do a reasonable job of photographing night scenes, though you may need to hold the device steady, ideally on a tripod, to avoid blurring. Daytime shutter speeds are usually around 1/200 second, but at night the shutter has to be open for longer to compensate for the low light level. To avoid blurring in hand-held shots, the longest shutter speed they offer is usually around ¼ second.

However, to be able to capture true night views, such as the stars, Milky Way and the aurora, much longer exposure times are needed. More advanced cameras will give these on the Program or Manual setting, but on most phone cameras you may need to buy an app to extend the range for real night photography.

For iPhones, NightCap is very popular, though be aware that its effectiveness depends on the iPhone you have. It will keep the shutter open for long periods, and also has specific modes for star trails and meteors. It isn't currently available for Android phones, though there are apps such as Open Camera which give you much more manual control over your phone's settings. And again, it won't work brilliantly with every camera.

Most phone cameras have only a wide-angle lens, because phones are thin and telephoto lenses need a longer distance between the lens and the light-sensitive chip. Some do have separate telephoto lenses, or you can buy add-on converters, but if you need longer telephoto lenses you might as well buy a separate camera anyway.

THROUGH THE TELESCOPE

What about taking pictures through the telescope? For bright objects such as the Moon and planets, this is perfectly feasible using a phone or compact camera, simply by pointing the camera into the telescope eyepiece – what's known as *afocal* photography. Getting everything in line can be tricky, but you can buy adapters for both cameras and phones – search online for digiscoping. Use a low-magnification, wide-field eyepiece and you should

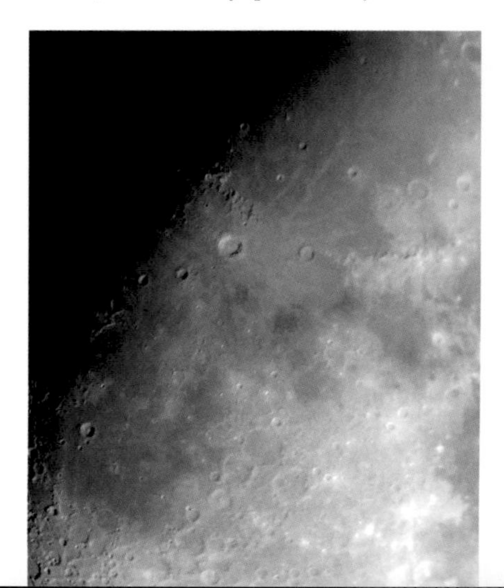

Cath Adams attached her iPhone 7 to her 114 mm reflector using an adapter and snapped this view of Copernicus and Mare Nubium. She used a 10-second delay to prevent image blur when operating the shutter.

be able to see the image on the camera or phone to check that the focus is OK.

One big problem with this is getting the exposure right. Unless the object fills the field of view the auto-exposure system on the camera sees mostly dark background and increases the exposure accordingly, so the object you're photographing is overexposed. If you can't control the exposure manually, one trick is to take the shot when there is still some light in the sky, so that the sky background is not too dark.

Afocal photography can often be a struggle, however, and has the drawback that you are photographing through both the eyepiece and the camera lens, so there is a fair bit of glass in the way which can introduce distortion of one sort or another. Ideally you need to be able to remove both the camera lens and eyepiece, which means using either a camera with removable lens such as a DSLR, or a specialist astronomy camera. Adapters are available for DSLRs, and with all but entry-level telescopes it's usually possible to get a focused image directly, with the telescope acting as a long telephoto lens.

The specialist astronomy cameras are based on webcam technology and provide a video stream that requires a laptop to capture the data over a period of a minute or two. Then you use free software such as RegiStax or AutoStakkert! to choose automatically the best frames and stack them so as to reduce the electronic noise in the image. Finally, sharpening software performs an almost magical transformation to bring out the fine detail.

The practicalities of this are beyond the scope of this article, but even a small telescope can yield excellent lunar and planetary images. In the case of the Moon,

It may not be in the Damian Peach class, but this photo of Jupiter and Ganymede was taken through a budget Sky-Watcher 130 mm reflector with a ZWO ASI120MC camera.

exposure times are similar to daytime photography, but for the planets some magnification is needed, usually provided by a Barlow lens of the sort used to amplify the power of an eyepiece. Ideally the telescope should be motor-driven so that the planet remains in the field of view while the video is being captured.

LONG EXPOSURES

But photographing bright planets isn't enough for many people – they want to capture the fainter objects, such as nebulae, clusters, galaxies and comets. This is where long time exposures, often minutes long, become essential. Equally crucial is some means of tracking the objects for this length of time as they move through the sky. While simple motor drives will keep a planet within the field of view for minutes at a time, to photograph stars you need very much better tracking.

The longer the focal length of the lens or telescope you're using, the more precise your tracking has to be. With a wide-angle lens, say the typical 18 mm setting of a DSLR zoom, you can give 30-second exposures without any tracking at all and the stars will not trail. But at 200 mm, anything longer than a second or two will require tracking, and when photographing through a telescope the requirements are more stringent still.

At this point, the type of mounting you use for your telescope becomes important. The simplest type is known as *altazimuth* or *altaz,* and has axes fixed vertically and horizontally. But objects moving through the sky usually move at an angle to the horizontal that varies depending where they are in the sky. Many telescopes these days are on this type of mount, with motor drives that vary the rate of each axis to follow objects smoothly. They are fine for photographing planets, where the object is small and exposure times are short.

The other type is the *equatorial,* which has axes that need to be tilted at an angle to the horizon so that objects can be tracked using one movement only. These need to be aligned quite carefully north–south, and at the precise angle according to your location – a task that defeats many beginners.

Track a star field containing a nebula or galaxy for any length of time with an altaz mount and, no matter how precise the tracking, very soon the field of view will start to rotate so stars at the edge become trails even if the centre remains steady. But with a well-aligned equatorial, everything remains fixed. You can sometimes get away with giving short exposures on an altazimuth mount and processing the images afterwards by stacking them with corrected rotation, but often this fails because of lens distortions across the field of view.

Another complication is that motor drives almost always have slight imperfections in their drive rate, known as *periodic error,* which mean that star trails wriggle. This can be corrected in various ways, the most reliable and popular these days being to have a small separate guide telescope on the mounting, with its own autoguider camera, whose job is to provide feedback to the mounting and give good tracking. So the idea of just fixing your camera looking through your telescope and giving a time exposure of a few minutes to record that nebula or galaxy rapidly becomes increasingly complicated and indeed costly, with power supplies, cables and devices all needing to be perfectly aligned and kept in sync. My advice is to not embark on this lightly, and to do a lot of research first. Having said that, the results can be rewarding, even for city dwellers,

The Sky-Watcher Skymax 102 is a sub-£400 Maksutov on a single-arm altazimuth mount that will allow you to start imaging the Moon, planets and a few deep-sky objects using lightweight astro cameras. It has Go To computer control which will locate and track a wide range of objects.

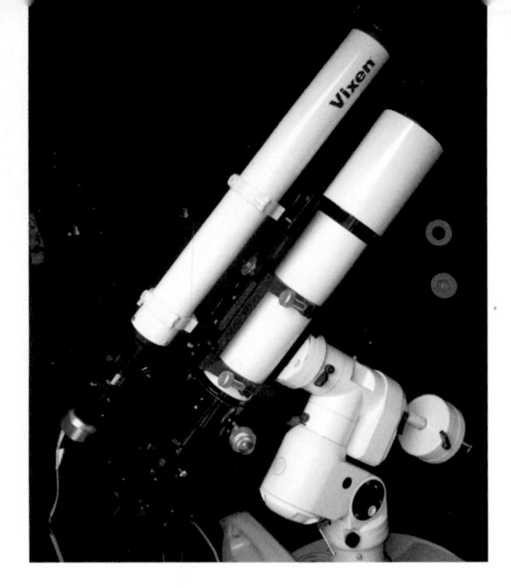

At the advanced end of the scale, Nick Hart in light-polluted Bridgend uses this Ikharos 102 mm ED refractor on a driven Vixen Atlux equatorial mount. His guidescope is a Vixen 70 mm refractor equipped with an Orion Starshoot autoguiding camera. On the main telescope is an Atik 383L CCD camera equipped with filter wheel.

as narrowband filters can help combat light pollution. The photo on page 59 by Simon Hudson, taken from within Greater London, is a great example.

CHOOSING YOUR EQUIPMENT

You can do some good astrophotography with a telescope in the aperture range 100 to 150 mm, and sometimes even smaller. Newtonian, Schmidt-Cassegrain or Maksutov reflecting telescopes usually work well, as do apochromatic or ED refractors. Expect to pay at least £400 for a reasonable instrument. For the Moon and planets a motorised altazimuth mount will do the job, and a specialist planetary camera such as one of the ZWO series will cost upwards of £200.

The standard everyday DSLR camera is perfectly adequate for much deep-sky work, particularly if your skies are dark so light pollution is not a problem. Telescopes as small as 60 mm or even good telephoto lenses will get you started on the larger galaxies and nebulae, but you'll need a motor-driven mount which will track the stars. John Bell's Hyades picture on page 11 shows what you can do.

For lightweight telescopes and telephoto lenses a tracking mount such as the Sky-Watcher Star Adventurer is fine, but heavier and longer focal-length telescopes require a larger equatorial mount, ideally with an autoguider port. We're talking of a starting price around £650, plus the cost of a camera and guide scope with autoguider camera for long-exposure imaging. Your DSLR will be a good start, but a good-quality CCD camera will set you back at least £1000. And you'll need a decent laptop for the image capture.

Fortunately, you don't need to splash out all at once, but if your heart is set on achieving beautiful deep-sky images you can certainly build yourself a set-up which will take colourful images of objects which until recent years could scarcely be seen from most UK locations.

With the above set-up Nick produced this splendid view of the Rosette Nebula, NGC 2244, imaging through separate red, green and blue colour filters plus a narrowband OIII filter.

Our view of the stars – a source of infinite amazement for scientists, stargazers and the millions of us who seek out rural places to rest and recuperate – is obscured by light pollution. It's a sad fact that many people may never see the Milky Way, our own Galaxy, because of the impact of artificial light.

LIGHT POLLUTION

Light pollution is a generic term referring to artificial light that shines where it is neither wanted nor needed. In broad terms, there are three types of light pollution:

- **Skyglow** – the pink or orange glow we see for miles around towns and cities, spreading deep into the countryside, caused by a scattering of artificial light by airborne dust and water droplets.
- **Glare** – the uncomfortable brightness of a light source.
- **Light intrusion** – light spilling beyond the boundary of the property on which a light is located, sometimes shining through windows and curtains.

CPRE, the countryside charity, has long fought for the protection and improvement of dark skies, and against the spread of unnecessary artificial light. CPRE commissioned LUC to create new maps of Great Britain's light pollution and dark skies to give an accurate picture of how much light is spilling up into the night sky and show where urgent action is needed. CPRE also sought to find where the darkest skies are, so that they can be protected and improved.

MAPPING

The maps are based on data gathered by the National Oceanographic and Atmospheric Administration (NOAA) in America, using the Suomi NPP weather satellite. One of the instruments on board the satellite is the Visible Infrared Imaging Radiometer Suite (VIIRS), which captures visible and infrared imagery to monitor and measure processes on Earth, including the amount of light spilling up into the night sky. This light is captured by a day/night band sensor.

The mapping used data gathered in September 2015, and is made up of a composite of nightly images taken that month as the satellite passed over the UK at 1.30 am.

The data was split into nine categories to distinguish between different light levels. Colours were assigned to each category, ranging from darkest to brightest, as shown in the chart below. The maps

Colour bandings to show levels of brightness

Categories	Brightness values (in nw/cm²/sr)*
Colour band 1 (darkest)	<0.25
Colour band 2	0.25–0.5
Colour band 3	0.5–1
Colour band 4	1–2
Colour band 5	2–4
Colour band 6	4–8
Colour band 7	8–16
Colour band 8	16–32
Colour band 9 (brightest)	>32

**The brightness values are measured in nanowatts/cm²/steradian (nw/cm²/sr). In simple terms, this calculates how the satellite instruments measure the light on the ground, taking account of the distance between the two.*

are divided into pixels, 400 metres × 400 metres, to show the amount of light shining up into the night sky from that area. This is measured by the satellite in nanowatts, which is then used to create a measure of night-time brightness.

The nine colour bands were applied to a national map of Great Britain (see the following pages), which clearly identifies the main concentrations of night-time lights, creating light pollution that spills up into the sky.

The highest levels of light pollution are around towns and cities, with the highest densities around London, Leeds, Manchester, Liverpool, Birmingham and Newcastle. Heavily lit transport infrastructure, such as major roads, ports and airports, also show up clearly on the map. The national map also shows that there are many areas that have very little light pollution, where people can expect to see a truly dark night sky.

The results show that only 21.7 per cent of England has pristine night skies, completely free from light pollution (see the chart below). This compares with almost 57 per cent of Wales and 77 per cent of Scotland. When the two darkest categories are combined, 49 per cent of England can be considered dark, compared with almost 75 per cent in Wales and 87.5 per cent in Scotland. There are noticeably higher levels of light pollution in England in all the categories, compared with Wales and Scotland. The amount of the most severe light pollution is five times higher in England than in Scotland and six times higher than in Wales.

The different levels of light pollution are linked to the varying population densities of the three countries: where there are higher population densities, there are higher levels of light pollution. For example, the Welsh Valleys are clearly shown by the fingers of light pollution spreading north from Newport, Cardiff, Bridgend and Swansea. In Scotland, the main populated areas stretching from Edinburgh to Glasgow show almost unbroken levels of light pollution, creeping out from the cities and towns to blur any distinction between urban and rural areas.

Light levels in England, Wales and Scotland

Categories	England	Wales	Scotland	GB
Colour band 1 (darkest)	21.7%	56.9%	76.8%	46.2%
Colour band 2	27.3%	18.0%	10.7%	20.1%
Colour band 3	19.1%	9.3%	4.6%	12.6%
Colour band 4	11.0%	5.8%	2.8%	7.3%
Colour band 5	6.8%	3.8%	1.7%	4.6%
Colour band 6	5.0%	2.9%	1.2%	3.3%
Colour band 7	4.3%	2.1%	1.0%	2.8%
Colour band 8	3.2%	1.0%	0.9%	2.1%
Colour band 9 (brightest)	1.6%	0.2%	0.3%	1.0%

Adapted from Night Blight: Mapping England's light pollution and dark skies *CPRE (2016), with kind permission from CPRE. To see the full report and dedicated website, go to http://nightblight.cpre.org.uk/*

MAP OF BRITAIN'S LIGHT POLLUTION AND DARK SKIES

COURTESY OF CPRE/LUC

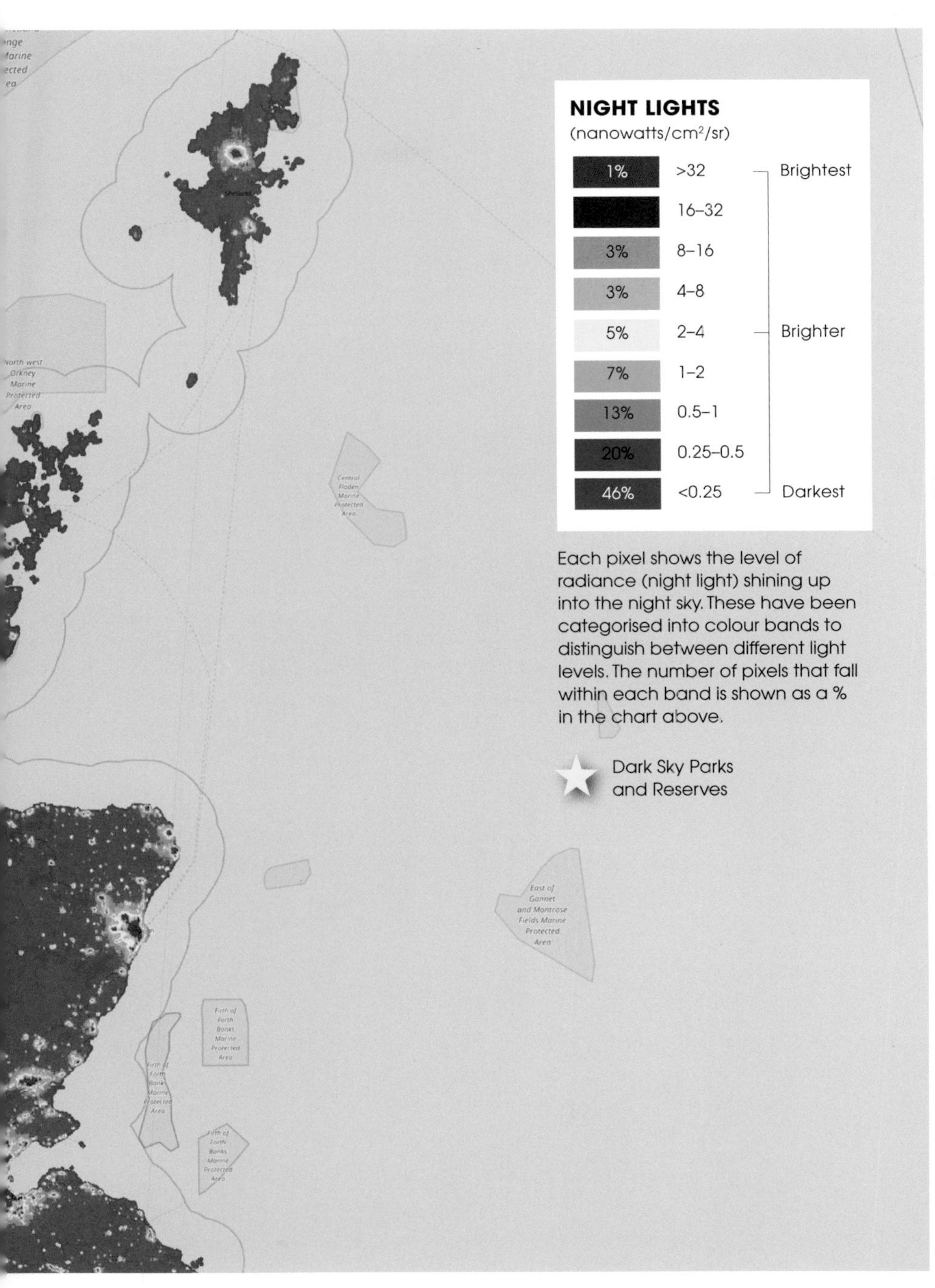

Each pixel shows the level of radiance (night light) shining up into the night sky. These have been categorised into colour bands to distinguish between different light levels. The number of pixels that fall within each band is shown as a % in the chart above.

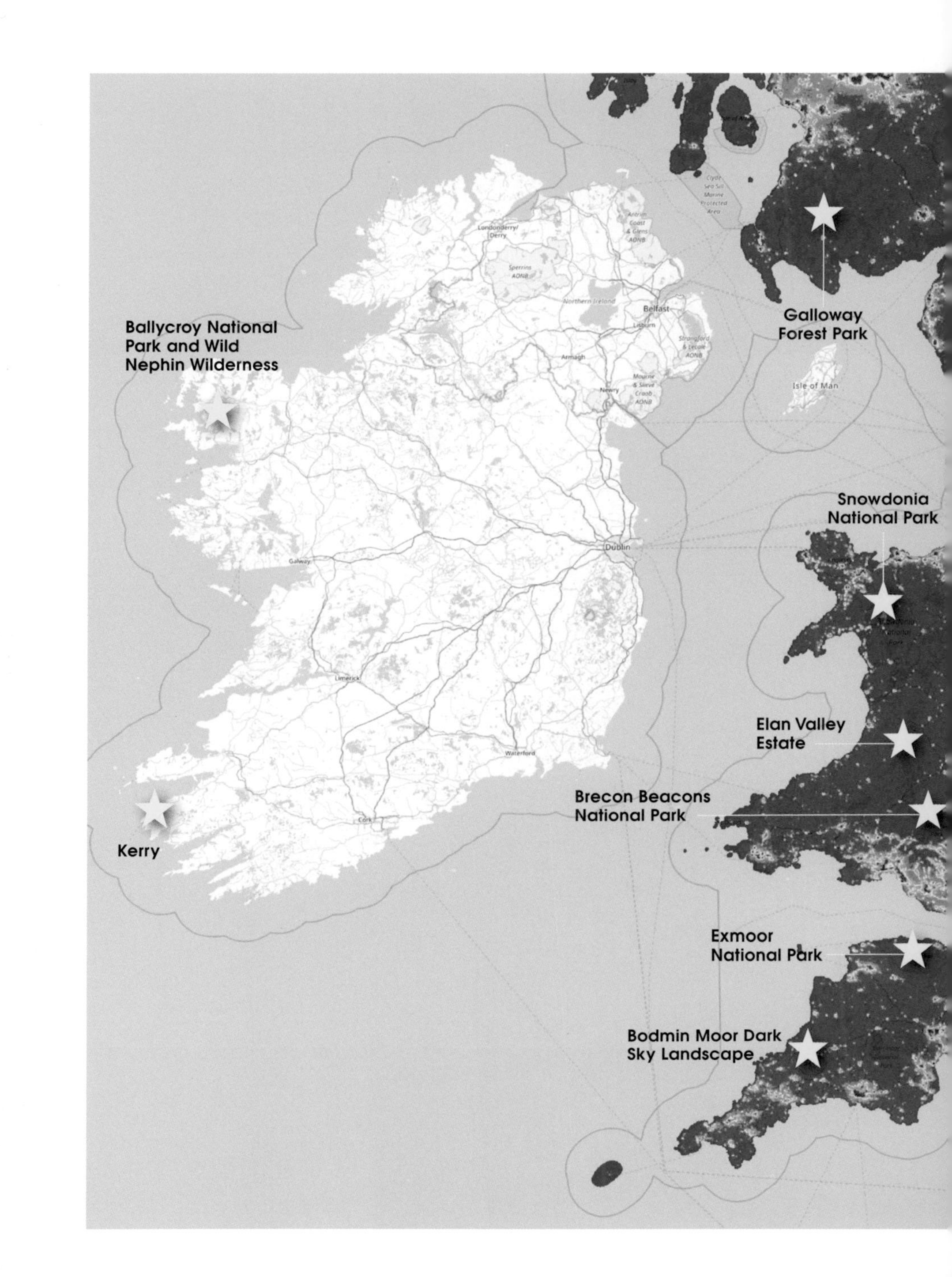
Ballycroy National Park and Wild Nephin Wilderness
Kerry
Galloway Forest Park
Snowdonia National Park
Elan Valley Estate
Brecon Beacons National Park
Exmoor National Park
Bodmin Moor Dark Sky Landscape
Londonderry/ Derry
Belfast
Lisburn
Armagh
Newry
Dublin
Galway
Limerick
Waterford
Cork
Isle of Man
Northern Ireland

Dark Sky Parks and Reserves
Moffat
Northumberland National Park and Kielder Water and Forest Park
Moore's Reserve (South Downs)
Brugge
Gent
Dunkerque
Calais
Lille
Mons
Zeeland

THE AUTHORS

Professors **Heather Couper** and **Nigel Henbest** are both Fellows of the Royal Astronomical Society. Recognised internationally as writers and broadcasters on astronomy and space, they have penned 50 books and over 1000 articles in newspapers and magazines.

They co-founded a highly successful TV production companion which has clocked up 600 hours of factual programmes for the international market.

Heather is past-President of the British Astronomical Association, and a Fellow of the Institute of Physics. She was awarded the CBE in 2007 for Services to Science.

After researching at Cambridge, Nigel became consultant to both *New Scientist* magazine and the Royal Greenwich Observatory. He is a future astronaut with Vir[illegible] Galactic.

ACKNOWLEDGEMENTS

PHOTOGRAPHS
Front cover: Damian Peach, Saturn.
Galaxy Picture Library: Cath Adams 86; John Bell 11; Nigel Bradbury 41; Jamie Cooper 34, 77; Nick Hart 85, 89 (top and bottom); Thomas Heaton 29; Craig Howman 23; Simon Hudson 59; Damian Peach and Sebastian Voltmer 71; Nils Reuther 17; Robin Scagell 6, 84 (top), 87; Peter Shah 83, 84 (bottom); Pete Williamson 1, 47, 65; Alastair Woodward 53.
NASA: /NSS-DCA 13; /JPL 37; Bob Franke 60; JPL-Caltech 67; Sean M. Sabatini 72; /JPL/University of Arizona 79.
Hencoup Enterprises: Nigel Henbest 7, 96.
Optical Vision Ltd: 88.

ARTWORKS
Star maps: Wil Tirion/Philip's with extra annotation by Philip's.
Planet event charts: Nigel Henbest/Stellarium (www.stellarium.org).
Pages 80–82: Chris Bell.
Pages 90–95: Adapted from *Night Blight: Mapping England's light pollution and dark skies* CPRE (2016), with kind permission from CPRE.
To see the full report and dedicated website, go to http://nightblight.cpre.org.uk/
Maps © OpenStreetMap contributors, Earth Observation Group, NOAA National Geophysical Data Center. Developed by CPRE and LUC.